EARTHWATCH

space-time investigations with a globe

by

JULIUS SCHWARTZ

illustrated by

RADU VERO

McGRAW-HILL BOOK COMPANY

New York St. Louis San Francisco Auckland Bogotá Düsseldorf Johannesburg London Madrid Mexico Montreal New Delhi Panama Paris São Paulo Singapore Sydney Tokyo Toronto

OTHER BOOKS BY THE AUTHOR

IT'S FUN TO KNOW WHY
THE EARTH IS YOUR SPACESHIP
THROUGH THE MAGNIFYING GLASS
MAGNIFY AND FIND OUT WHY

Library of Congress Cataloging in Publication Data

Schwartz, Julius, date-
Earthwatch: space-time investigations with a globe.

Bibliography: p.
Includes index.
SUMMARY: Includes a variety of experiments with a globe to illustrate the movement of the earth and its satellites.
1. Earth—Juvenile literature. 2. Astronomy, Spherical and practical—Juvenile literature.
3. Globes—Juvenile literature. [1. Earth.
2. Astronomy. 3. Globes] I. Vero, R.
II. Title.
QB631.S34 525 76-30334
ISBN 0-07-055685-7 lib. bdg.

12345 RAHO 78987

Contents

For my grandson, Jean-Paul

The Earth is in your hands!

Acknowledgements

To the late Richard M. Sutton, Professor of Physics, California Institute of Technology, for his demonstration of the "rectified globe" as an operating model of the Earth in space.

To Fletcher G. Watson, Professor of Education, Graduate School of Education, Harvard University, for his critical review of the original manuscript and for his many helpful suggestions with respect to the techniques employed and the concepts presented.

Of course the author alone accepts responsibility for the final work.

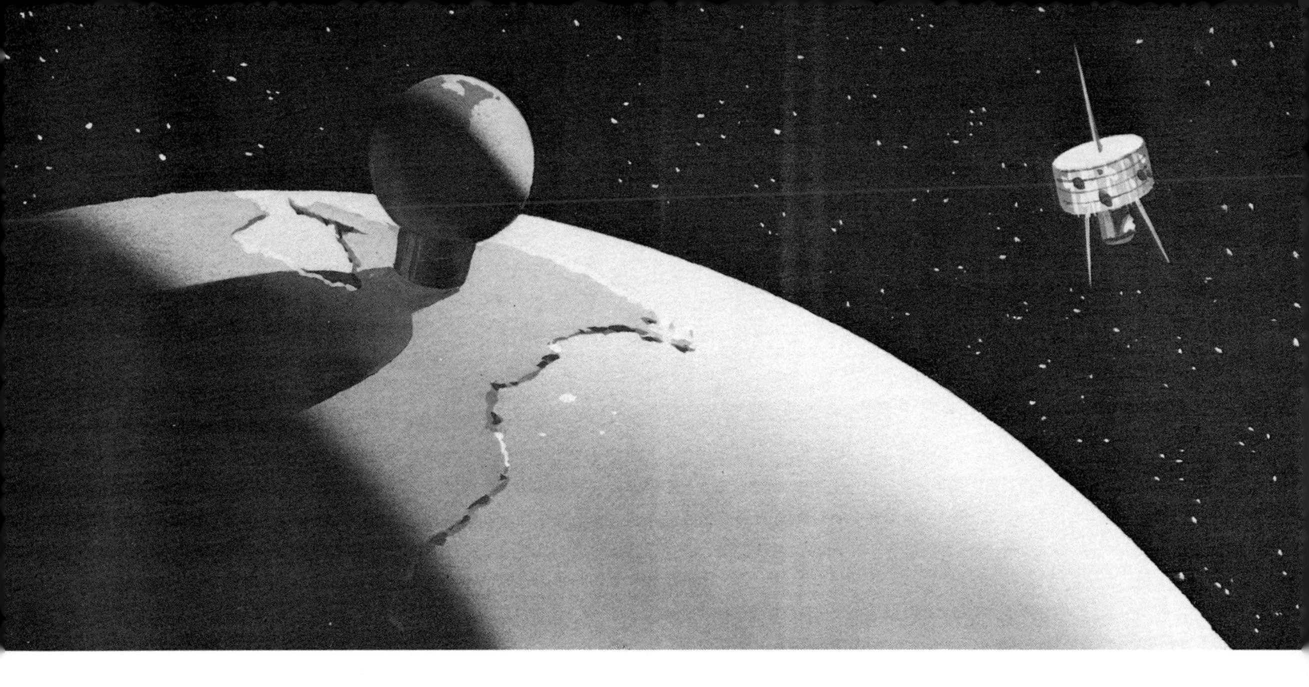

Chapter I-Your Own Satellite

More than a thousand satellites accompany the Earth in its journey through space. As they orbit our planet in their assigned tracks they serve many useful purposes. Meteorological satellites photograph and transmit to Earth the cloud pictures seen on television weather reports. Astronomical spacecraft high above the dust and turbulence of the Earth's atmosphere point precisely and steadily at the sun and distant stars. Communication satellites make possible live television broadcasts around the Earth. With the aid of Earth Resources satellites we are beginning to monitor and better manage the land, water, and mineral wealth of the Earth. And these are but a few of the many services provided by the fleet of satellites attending the Earth.

An ordinary world globe can serve as your own monitor of our planet—if it is mounted properly on its "pad" in sunlight. The globe then becomes a kind of satellite—not one that is rocketed into space, but instead one that remains attached to its home planet. It is this permanent connection that makes the globe an accurate reporter of what the Earth is doing in space, minute after minute, day after day, season after season.

With such a globe you can see where sunrise and sunset are sweeping across the Earth; locate the "hot spot" on Earth where the sun is directly overhead; feel where on Earth it is hot, warm, or cool; find out how fast you are spinning in space; see the "land of the midnight sun"; view the Earth as if you were on the moon; count the number of hours of sunlight for any place on Earth; determine the Earth's position in its annual orbit around the sun; and observe many other space happenings on our planet.

Properly positioned in sunlight, the geographic globe becomes an instrument for investigating planet Earth's journey through time and space. With it you can conduct your own "Project Earthwatch."

Fig. 1. Tilt the globe so that the place where you live is on top.

Fig. 2. Turn the globe so that its axis points to the north.

Chapter II-The Globe on its Space Pad

To work as a space monitor, your globe should be in direct sunlight for as many hours a day as possible. An open area outdoors, away from the shadows of buildings or trees is best, even if you have to carry your globe there each time you wish to use it. A sunlit room facing south is good—and in some respects better than the outdoors, as you will see later.

But it is not enough to place the globe in sunlight. If it is to bring news of the Earth it must be tilted and turned so that it has the same orientation in space as the Earth itself. Like a dancer in a ballet executing the same movement at the same moment as another dancer, the globe must have a space posture identical to that of the Earth. This is easily accomplished in two steps.

TILT AND TURN

TILT: *Tilt the globe so that the place where you live—or the city on the globe nearest you—is on top.* To keep it in that position you may have to lift the globe and mount it on some kind of support such as a bowl, as shown in the illustration (Fig. 1). (You may find this easier if you first "free" the globe by unscrewing it from its original mounting.) Or, perhaps you have the kind of globe that is built to tilt in any position. To make sure that your place is on top, stand a chess piece, a golf tee, a suction cup "arrow," or any other similar object on your home location. Step back from the globe, view it from different directions, and adjust it if necessary.

TURN: The second step is a bit more difficult; at least, it may be so the *first* time you do it. *The axis of the globe must be turned so that it points to the North.* The axis is the imaginary line or the actual rod going through the globe from the South Pole through the North Pole. While you are turning the globe to point it properly you must *keep your home place on top* (Fig. 2).

You may find it easier to use the lines on the globe that connect the poles, rather than the axis. Find the one near where you live and adjust the globe until this line runs in a true north-south direction. If you live some distance from such a line, make your own by connecting your place to the North Pole with a string, pencil line, or strip of paper.

But which way is north?

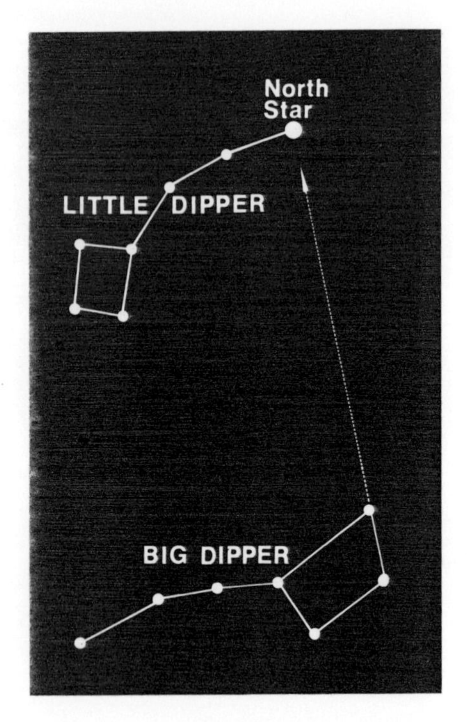

Fig. 3. A line through the "pointers" leads to the North Star.

FINDING NORTH

There are a number of ways of finding north. Perhaps you know where north is on your street. Or you may have a map of your town that shows north. Perhaps you have found—or can find—the North Star at night. The star map in Fig. 3 shows how. Or you may use a compass to help you—but compass north where you live is not necessarily the same as true north. In that case you must make a correction for what is called the compass declination. The illustration shows how to do this (Fig. 5). Notice that the correction is different in amount and direction for different places in the country.

Another way of finding north is to use a shadow stick. Attach a stick to a flat board as shown in the drawing. Place the board in a level position outdoors in sunlight and watch the shadow of the stick. In the morning, draw a line over the shadow to show its position and length. In the afternoon, when the shadow is equally long, draw a second line over it. Divide in half the angle formed by the morning and

Fig. 4. A shadow stick helps you find north.

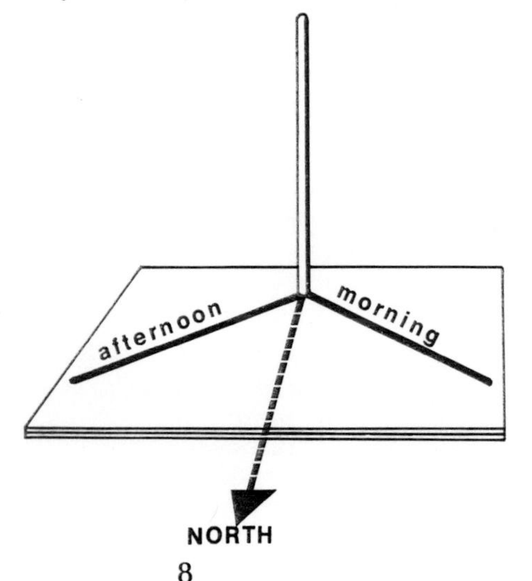

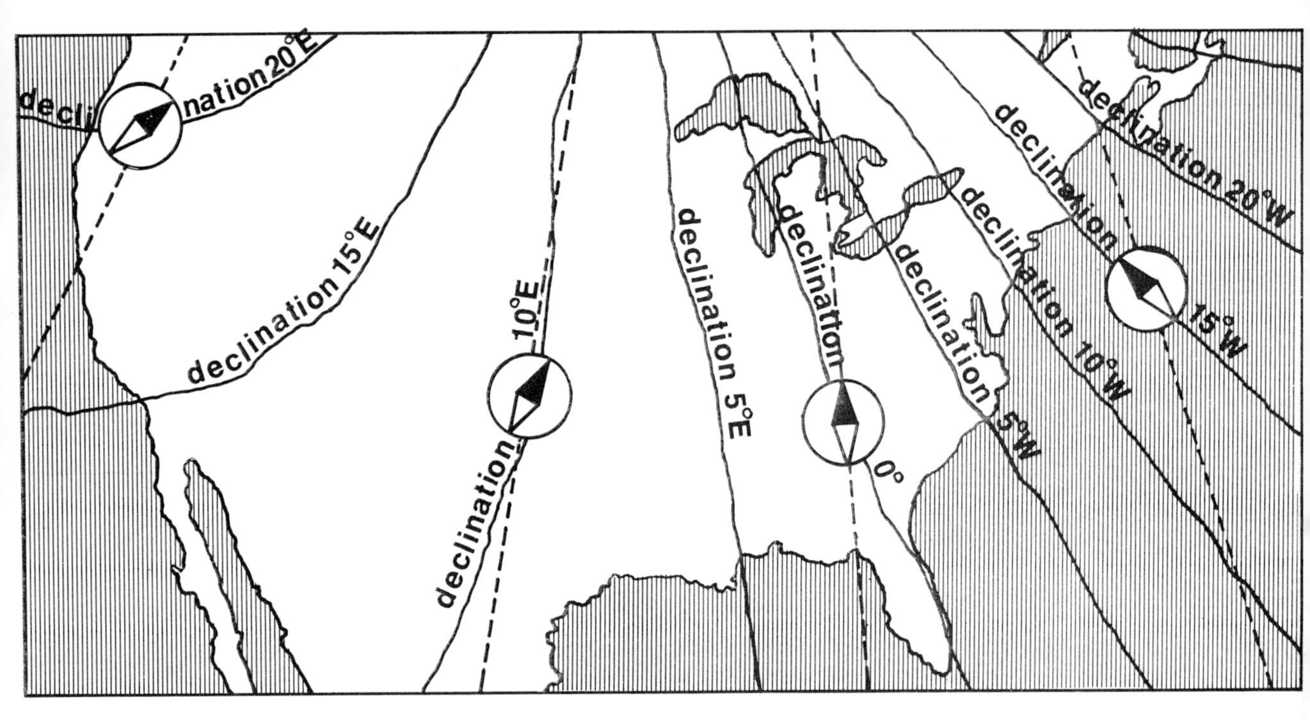

Fig. 5. How to correct your compass. The broken lines represent true or geographic north. The solid lines indicate the difference between true north and compass north—the declination. For example: If you are located along the line 10° E, your compass will point 10° east of true north. Consequently, true north is 10° west of the compass reading.

afternoon shadows. This third line points northward (Fig. 4).

Wherever you are, it is a good idea to make note of the north-south direction. One way to do this is to find two fixed points—a tree and the corner of a house, for example—that line up in a north-south direction. Another way is to draw a north-south line on the pavement or to drive two posts into the ground, if that is permitted. This will make it easier to position the globe properly each time you use it (Fig. 6).

THE SPACE-POSITIONED GLOBE

Consider what you have accomplished by mounting the globe in its space position (Fig. 7):

On the globe the place where you live is on top.
(On Earth, wherever you live, you are also "on top.")
The axis of the globe parallels the Earth's axis, which points near the North Star.

Your globe is now a model Earth that shows what big Earth is doing in space. Although it remains on its space pad on Earth, it is a useful satellite, ready for its job.

Fig. 6. Two objects help fix the north-south direction.

GLOBE CARE

A few words about the care of globes. It is best not to use an antique or rare globe, since it may get some rough wear at times. Also, when you've completed your observations for the day, bring the globe indoors. It is easy to set it up again, after the first time. At the conclusion of an investigation use a damp sponge or soft eraser to remove any marks you've made on the globe.

For a full understanding of what your space-mounted globe is telling you, first become acquainted with all of its parts. The following chapter will help.

Fig. 7. You are "on the top of the world" wherever you live.

Chapter III - Know Your Instrument

You have mounted your globe, your mini-earth, on its "space pad" in such a way that it will report accurately and immediately happenings on planet Earth. To "read" these reports with understanding it is important to become acquainted with the information printed on the globe. If, however, all this is familiar, skip through this chapter and come back later when necessary.

TWO POINTS AND A CIRCLE

On looking at a geographic globe, one's attention is first drawn to the large land masses and the surrounding oceans. A closer view reveals individual continents, countries, and cities, then oceanic islands and currents and, on some globes, under-ocean basins and ridges. For many, the globe's fascination is that it is a detailed model of the Earth itself.

Globes come in different sizes and colors, but all are spherical. A sphere has no beginning and no end. The Earth, however, is a sphere that *turns*. So two turning points are marked on the surface of the globe, the **poles**. Midway between the poles a circle is drawn around the globe, the **equator** (Fig. 8). Find the poles and the equator on your globe.

The North Pole and South Pole are exactly opposite each other on the sphere, at the ends of the imaginary **axis** on which the Earth turns. If the axis

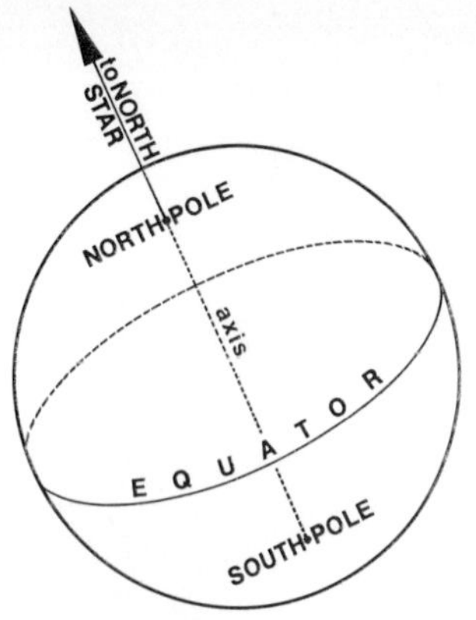

Fig. 8. The poles are at the ends of the Earth's axis.

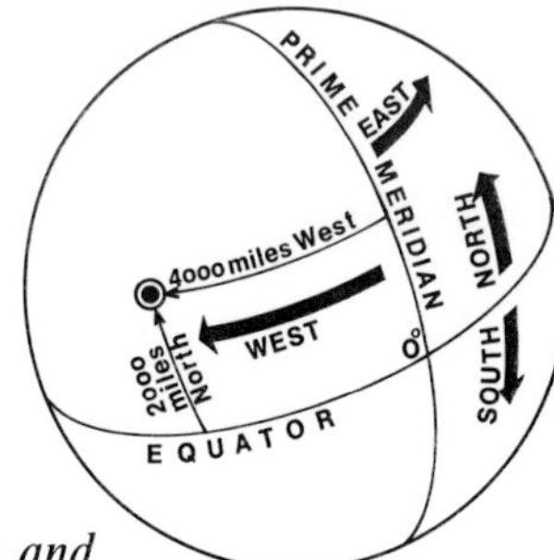

Fig. 9. The equator and the prime meridian fix the spot.

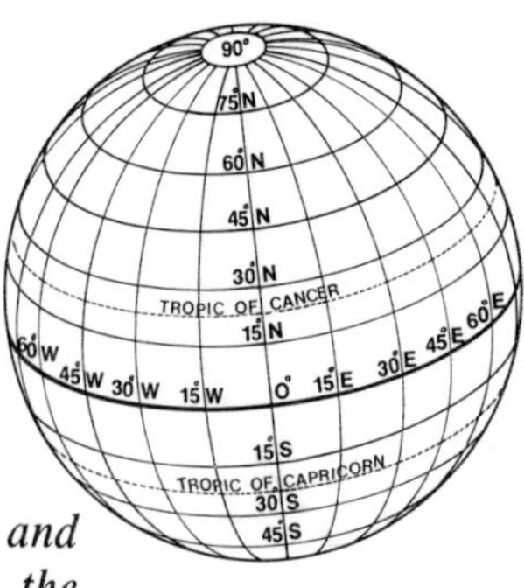

Fig. 10. The parallels and meridians are like the streets and avenues in your town.

Fig. 11. Going places.

line were continued out into space from the North Pole, it would pass near the North Star.

The equator divides the surface of the globe into two equal halves, the Northern and the Southern Hemispheres. Run your finger around the equator and you trace a **great circle**, the largest possible circle that one may draw on a sphere. An infinite number of great circles may be drawn on a sphere, but on the globe only one, the equator, is equally distant from both poles.

A LOCATION SYSTEM

The equator and the poles can be used to build a location system that will pinpoint any object or event on the Earth. For example, where is the eye of a hurricane right now? Where will splashdown be for a spacecraft returning to Earth? Where will the moon's shadow touch the Earth during an eclipse of the sun?

The equator is a natural baseline to begin with. We could say, for example, that the place we wish to locate is 3,200 kilometers (2,000 miles) north of the equator (Fig. 9). Of course that would not tell us enough since there is a line of points 3,200 kilometers north of the equator. Together the points form a circle around the globe 3,200 kilometers north of the equator and parallel to it. Which point is the right one?

So another baseline had to be invented from which one could measure distances east and west along the circle. Such a line would have to run north-south. Where should it be drawn?

By agreement of the countries participating in the Washington Meridian Conference of 1884, a line called the **prime meridian** ("prime" from the Latin, meaning "first" or "most im-

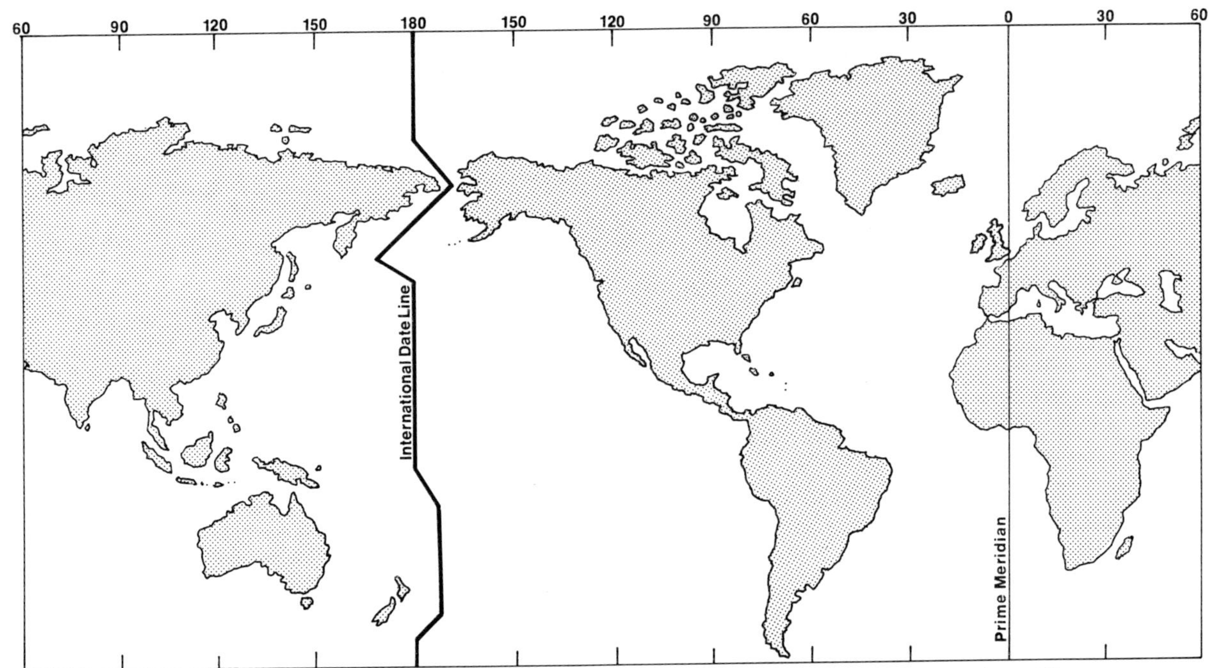

Fig. 12. There are always two days going on the Earth at the same time. The 180° meridian, the international date line, divides the new day from the old. If you cross it going toward Asia on Monday, you fly or sail into Tuesday. Go the opposite way and you'll go from Tuesday into Monday. The old day is always on one side of the date line and the new one on the other.

portant") was designated. The prime meridian is half of a great circle that extends from the North Pole through Greenwich, England, a town near London, to the South Pole. Find the prime meridian on your globe.

Now it would be possible to say: The place we wish to pinpoint is 3,200 kilometers north of the equator and 6,400 kilometers (4,000 miles) west of the prime meridian. That would fix the spot.

In addition to the equator and the prime meridian, a network of crossing lines is found on the globe. The lines running east and west are parallel to the equator. They are called **parallels**. The north-south lines connecting the poles are called **meridians** (Fig. 10).

Except for the equator, each of the parallels are **small circles**; they vary in circumference, becoming smaller and smaller from the equator to the poles.

All the meridians, on the other hand, are equal in length. Each is half of a great circle.

Any two parallels are equally distant from each other all the way around the globe. Meridians, however, are farthest apart at the equator and come closer and closer together toward the poles, where they meet in a point.

GOING PLACES

A few words about direction (Fig. 11). "Going north" means moving toward a definite *point* on the Earth's spherical surface—toward the North Pole.

When you reach the North Pole there is only one direction you can go (provided you don't blast off into space) and that is south. "Going south" means heading toward the South Pole.

East and west, on the other hand, are continuous directions. If you travel east far enough, you will come back to where you started and you can keep on going eastward forever. The same is true for traveling westward.

A MATTER OF DEGREE

To pinpoint a place on the Earth, as we have found, we need only know its position in relation to the equator and the prime meridian, measured in miles or kilometers. However, there is another way of describing position on a globe.

Consider, for example, the location of New Orleans, Louisiana. Find the city on the globe and place the tip of the forefinger of each hand as close as possible to each other on it. Now slide both fingers at the same time: the left one on a meridian down to the equator—stop when you get there; the right one on a parallel eastward until you reach the prime meridian. Consider these two positions as your starting points.

Now move the fingers in the opposite direction from the original motion until they meet in New Orleans. Notice that each finger travels on a curved path: *a part of a circle* (Fig. 13).

Circles—great and small—are the guidelines on a globe. A part of a circle can be measured as a fraction of it: one-half of it, one-quarter of it, and so on. Long ago the ancients divided the circle into 360 equal parts. Each part, or **degree**, was 1/360 of the total. The whole circle has 360 degrees, written 360°.

From the equator to New Orleans is 1/12 of the great circle that passes northward through this city around the Earth; 1/12 of 360° is 30°. Find the east-west circle 30° north from the equator that passes through New Orleans.

From the prime meridian to New Orleans is one-quarter of a circle, or 90°. Find 90° on New Orleans' meridian; look for this numeral where the meridian crosses the equator.

New Orleans is 30° north of the equator and 90° east of the prime meridian.

Degrees are a convenient and accurate way of locating points on the Earth.

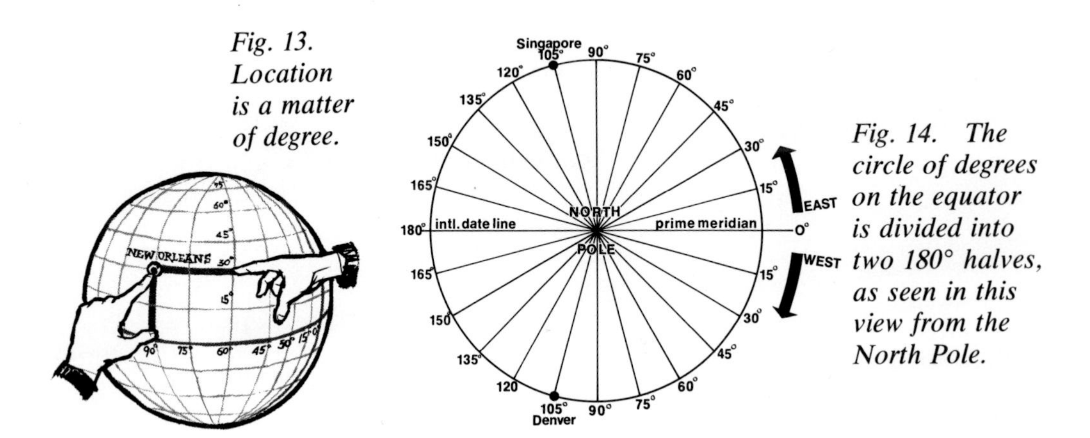

Fig. 13. Location is a matter of degree.

Fig. 14. The circle of degrees on the equator is divided into two 180° halves, as seen in this view from the North Pole.

HOW FAR NORTH? HOW FAR SOUTH?

Parallels are used to measure distances north and south of the equator. Find the parallel near where you live. Trace it around the globe until you find a numeral on it. If, for example, you live in Tallahassee, Florida, the figure is 30°. Now travel north to the next parallel. What is the figure there? On many globes the interval between parallels is 15°. In that case the answer would be 45°. (On other globes where there are 10° degree intervals the answer, of course, would be 40°.) Continuing north, the numerals advance for each parallel: 60°, 75°. The North Pole is a point 90° from the equator which may or may not be labeled 90° on your globe.

Traveling southward from the 30° parallel, you pass the 15° parallel and then come to the equator, the 0° parallel. South of the equator, as you might expect, the degrees advance from 0 to 90°. The parallels on the globe are used to indicate **latitude**, the distance in degrees north or south of the equator. To distinguish between Tallahassee and Grafton, Australia, which is also on a 30° parallel, we say that Tallahassee is at (approximately) 30° north latitude and Grafton at (approximately) 30° south latitude.

Ninety degrees north of the equator plus 90° south of the equator adds up to 180° in all. This is right because if you travel from the North to the South Pole you cover one-half of a circle or 180°.

In addition to the equally spaced parallels, many globes show four special ones. Find the Tropic of Cancer and the Arctic Circle north of the equator and the Tropic of Capricorn and the Antarctic Circle south of the equator. Estimate the latitude of each of these parallels. Later, in Chapter 11, it will be apparent why these parallels are significant in describing the journey of the Earth around the sun.

HOW FAR EAST? HOW FAR WEST?

Now look at the meridians, the north-south lines that cut across the parallels. Begin with the prime meridian and follow it down to the equator. Here you will find the figure 0° assigned to this meridian. Now slowly turn the globe to the right—toward the east. (This is the way the Earth turns.) Note that the numerals assigned to the meridians increase. On most globes the meridians are spaced 15° apart. Continuing to turn the globe, you reach a meridian labeled 180° that passes near the Fiji Islands. Along this meridian, but zigzagging in and out of it here and there, is the **international date line** (Fig. 12).

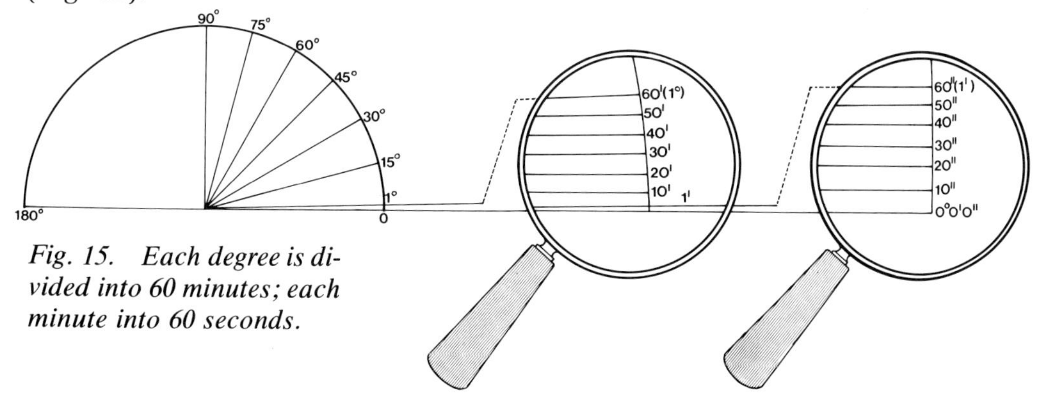

Fig. 15. Each degree is divided into 60 minutes; each minute into 60 seconds.

Fig. 16. Measuring the Earth's circumference.

Fig. 17. Measuring the distance between two cities.

If you cross this line while traveling eastward, the date is set back one day—Tuesday becomes Monday. The date is set forward one day when crossing the dateline going west—Monday becomes Tuesday. The international date line is kinked here and there so that it lies entirely in the ocean. In this way all changes of date are made on ship or plane; no body of land is divided by it.

Eastward from the international date line, the numbers assigned to the meridians decrease. Return to the prime meridian, but this time turn the globe toward the west. Again the figures (shown on the equator) for the meridians go up—from 0° to 180°.

Evidently, the degree circle marked on the equator for the meridians is divided into two 180° halves. In this sense it is different from that of a compass dial, which goes from 0° to 360°. Meridians are used to indicate **longitude**, degrees east or west of the prime meridian (Fig. 14).

How do we distinguish between two meridians with the same degree value? For example, one meridian marked 105° passes near Denver; another one also marked 105° goes through Singapore.

Any meridian west of the prime meridian up to 180° is considered to be a west meridian. Similarly, any meridian east of the prime meridian up to 180° is an east meridian. We say that Denver is (approximately) at 105° west longitude, or 105° W. Singapore is at 105° east longitude, or 105° E. (Fig. 14).

There is only one 0° meridian (the prime meridian) and one 180° meridian.

GETTING DOWN TO FINE POINTS

Latitude and longitude together make it possible to pinpoint any place on the globe (Fig. 13). New Orleans, Louisiana, for example, is located approximately at 30° north latitude and approximately 90° west longitude.

To make location finding more exact, each degree is divided into sixty smaller divisions called **minutes** (Fig. 15). New Orleans, for example, is at 29° 57′ (29 degrees, 57 minutes) north latitude and 90° 04′ (90 degrees, 4 minutes) west longitude.

For precise location even the subdivision of degrees into minutes is not fine enough. Each minute is divided into 60 seconds, written as 60″. For instance, the Washington Monument is located at 38° 53′ 21.681″ north latitude and 77° 02′ 07.955″ west longitude.

What is the latitude and longitude of your home town? A detailed map of your state or county may help you.

SCALE IT DOWN

There may be additional information on your globe about the scale used and the meaning of the different symbols. For the purpose of this book the following are important:

1. The diameter of the globe. A common size is 12 inches or 30.48 centimeters.

2. The natural scale. On a 12-inch globe it is in the ratio of 1 to 41,849,600 or 660.5 miles per inch (418.5 kilometers per centimeter).

You may use the scale to measure distances on the globe. For example, to find the circumference of the equator, wrap a string around it until it meets itself (Fig. 16). Lay out the string along a ruler, measure its length in inches, then multiply by the scale factor (660.5 on a 12-inch globe). If your answer is near 25,000 miles, you are close to right (the more exact figure is 24,901.5). If you are using centimeters to measure the circumference, you should come close to 40,000 kilometers (the more exact figure is 40,075).

To measure the distance between any two points on a globe, such as two cities, hold the string taut between them (Fig. 17), then lay out the string on a ruler and multiply as before. You may be surprised at some of the routes traced by a string used in this way. For example, the shortest route from Boston to Seoul, South Korea, is across the Arctic—not the Pacific Ocean.

A string when stretched on a globe follows a great circle path—the shortest distance possible between any two points on the surface of a sphere. This may sound strange since the great circle was described earlier as the largest circle one could draw on a sphere. To convince yourself, stretch a string between Halifax, Nova Scotia, and Salem, Oregon, both of which lie approximately on the 45° parallel (Fig. 18). Which is shorter: the small circle route along the parallel or the great circle route indicated by your string?

Do you see why this concept of a great circle route is helpful in navigation on the oceans and in the air?

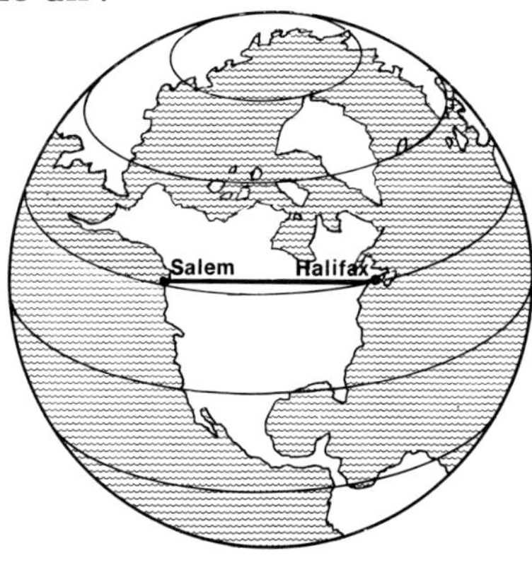

Fig. 18. A straight line is not the shortest distance between two places on the surface of a sphere.

Chapter IV-The Edge of Night

The Earth is a sunlit sphere in space. At any moment, however, only one-half of its surface is in sunlight. The globe on its space pad shows just where that half is. With your finger track the boundary line between light and dark. Your moving finger should be tracing a great circle that on Earth would be 25,000 miles around. Wander around the sunny side of the globe with your eyes and your fingers. What continents are found there? Oceans? Cities? How about the poles? Are both in sunlight?

The boundary line divides the Earth into two halves—the sunlit, daytime half; and the dark, nighttime half. You won't find it drawn on maps and globes because it changes minute by minute, day by day, season by season. Astronomers call the light-dark boundary the **terminator** ("termin" means "end") (Fig. 19). Every time you look at the moon when it is not full, you see *its* terminator marking the edge of its lighted surface. Draw the terminator on your globe with a soft pencil or crayon and record the time. Later the terminator will have moved westward.

Your globe, like the Earth, is a sunlit sphere in space. And since you have it on its space pad you can use it to find the edge of night circling the Earth. Later, as your globe turns with the Earth, you'll witness how cities and countries along half of the circle slip into the shadow of night, while places along the other half emerge into sunlight.

It doesn't matter where on Earth the space-positioned globe is viewed in sunlight: The sunlit and the dark halves of the globe will be the same. Observers in San Francisco, New York, Mexico City, Anchorage, and Buenos Aires will see the edge of night in the same place at the same time on their globes.

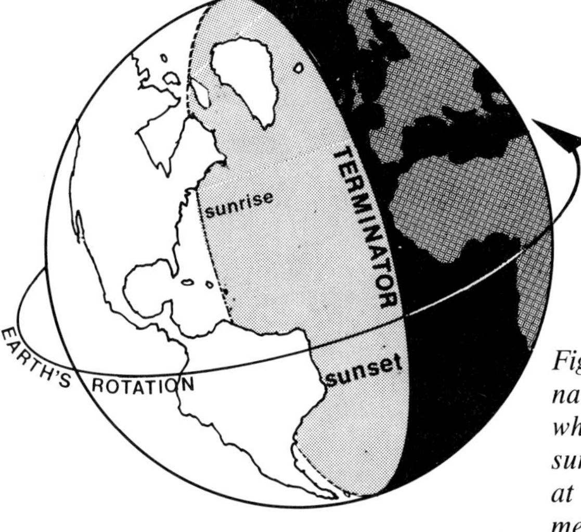

Fig. 19. The terminator marks the place where sunrise and sunset are occurring at a particular moment.

THE REFLECTED LIGHT PROBLEM

Sometimes it is hard to see the terminator clearly on your globe. Six ways of improving the seeing are:

1. Bring the globe indoors and set it up near a sunlit window. Then follow the TILT and TURN steps on pages 6-8.
2. Walk away from your globe and observe it from a distance.
3. Look at your globe when the sun is low in the sky, in early morning or late afternoon.
4. Look at the globe through orange or red sunglasses.
5. Build and use a shadow box from a carton that has been painted black on the inside, as shown in Figure 20.

Fig. 20. A shadow box excludes unwanted light.

6. Hold a pencil, eraser end down, on the sunlit side of the globe. Watch its shadow as you slide it slowly over the globe's surface. Where the shadow disappears is the darkness line you are looking for (Fig. 21).

Just why is it sometimes difficult to see the darkness line on your sunlit globe? For one thing, sunlight is reflected to the "night" side of the globe from the ground, nearby walls, and other surfaces. Perhaps more importantly, sky light also brightens this side.

The light of the daytime sky is a product of sunlight and air. The Earth is covered by an atmosphere a few hundred miles thick. The sun's light goes through the atmosphere and illuminates the surface of the Earth. On the way down some of the rays are scattered around and light up what we call the sky. Because of this the globe has its "night" side partly lit up by sky light. This may make it difficult to see the boundary line between its night and day halves. On the moon, which also is bathed in sunlight, you would not have a sky light problem with a lunar globe—because there is no atmosphere there to illuminate the sky. On the moon the sky is black and the stars are out even when the sun is shining.

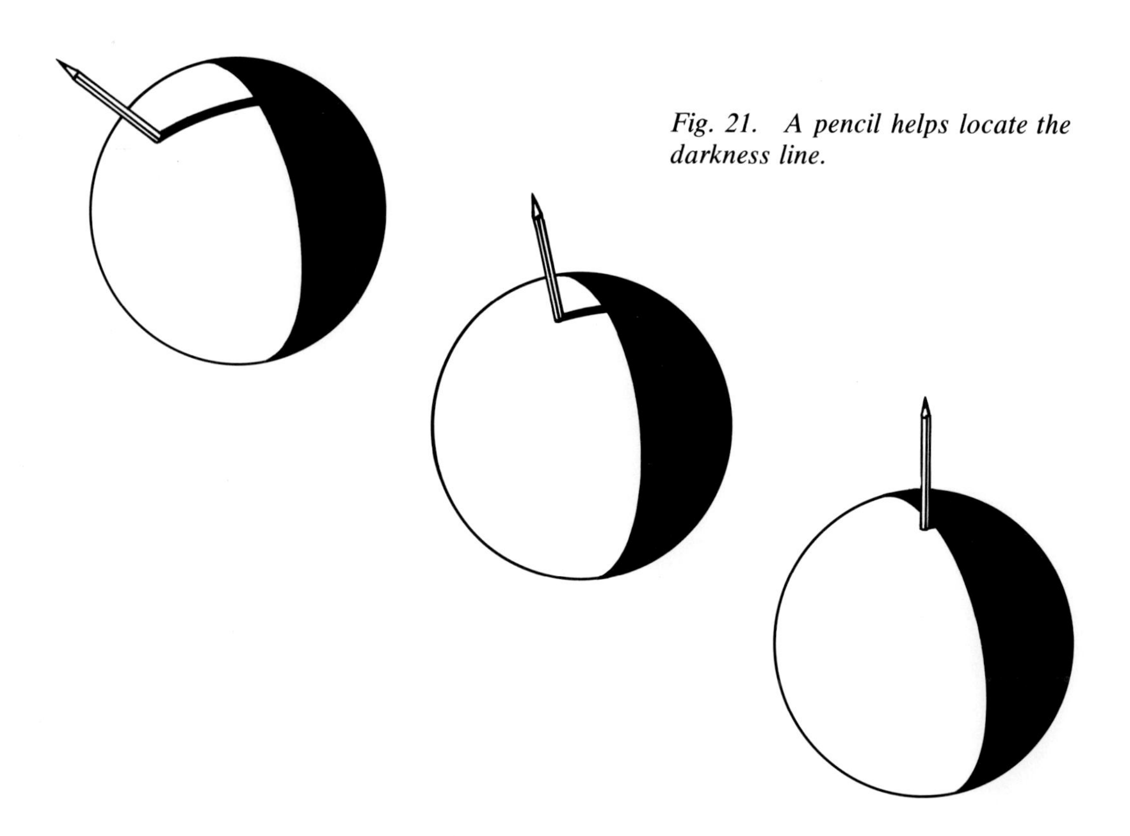

Fig. 21. A pencil helps locate the darkness line.

Chapter V-Sunset and Sunrise

In the late afternoon when the sun is low in the western sky, watch sunset creep over the globe. Find the sunset line to the east of where you live. Take a finger trip along the sunset line. What cities are having sunset right now? Place a thin strip of masking tape along this line. Instead of tape you may mark the globe with a soft pencil. While waiting for *your* sunset, follow the sunset line around the globe. On the opposite side it becomes the sunrise line! What cities are now having sunrise? Mark this place also with masking tape or a soft pencil.

Return to your globe in one hour. Where is the sunset line now? The sunrise line? Make a new mark for each on the globe (Fig. 22). What new places have just slipped into night? What others have moved into day? Both sunrise and sunset are always occurring—but for places on opposite sides of the Earth.

WHERE YOU LIVE

When the sun is close to the western horizon of the real Earth around you, what does the globe show? Is your town near the sunset line? Watch sunset on the Earth and on the globe. Look at your watch and note the time when the sun disappears. Compare this with the time given in your local newspaper.

Fig. 22. The strips mark the advance of sunset on the globe.
Fig. 23. Sunset sweeps westward as the Earth turns toward the east.

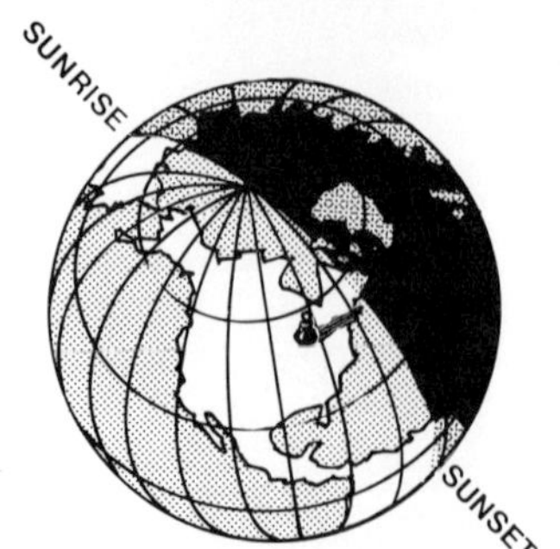

If you get up early enough, watch sunrise both on the real Earth and on your global model (Fig. 23). Observe how sunlight first touches the globe and then moves over it. Mark the progress of sunrise hour by hour as city after city west of your meridian passes across the edge of night into sunlight.

People invented the words sunrise and sunset because they observed the sun rising in the east, moving across the sky, and setting in the west. Now, of course, we know that it is the west-to-east rotation of the Earth that makes the sun appear to move.

THE EARTH TURNS THE GLOBE

What makes your globe "work"? What makes it an accurate reporter of Earth happenings? There is, after all, no secret clock hidden inside.

You know that your globe has the same orientation in space as the Earth and that it is lighted by the same sun. But you may still wonder what makes sunrise and sunset come to any place on the globe day after day at exactly the same time as on the Earth itself.

As the real Earth turns on its axis so does everything on it. Tennis balls and baseballs in your home, unless disturbed, rotate once a day, seven days a week. To understand how this occurs, picture yourself on a merry-go-round that represents the Earth. Assume that you are facing north at the outset and that the merry-go-round turns counterclockwise, viewed from above, or from left to right, as seen by a passerby on the ground. As the merry-go-round turns, you successively face buildings, trees, and other structures on the north, west, south, and east. This would be impossible unless your body made a complete turn on its top-to-toe axis—which it does (Fig. 24).

Similarly, the Earth turns your globe from west to east (counterclockwise when viewed from "above") and makes it "work."

Fig. 24. With each turn of the merry-go-round, the riders make a complete turn on their top-to-toe axes.

Chapter VI-View from the Moon

Picture a space ship leaving Earth, heading for the moon. How would the Earth look to the lunar astronauts during the 240,000 mile journey and later, after the moon landing? With the help of a globe you can stage a "voyage" to the moon and look back at the "Earth."

Mount your globe on its pad in sunlight. Since this is to be a space trip, you will need a large clear area around the globe. A lawn, a playground, a field, or a beach will do. The best blast-off time is in the early morning or late afternoon, when the sun is low in the sky.

THIRTY EARTHS AWAY

To see the Earth in its proper size from the moon you must journey outward from the globe. How far? It is helpful if you are guided by this relationship: *The moon is about thirty earth's diameters away from the Earth.* We arrive at the figure 30 by dividing the distance to the moon, 240,000 miles, by the diameter of the Earth, or 8,000 miles (Fig. 25).

Whatever the size of your globe, you will travel thirty globe diameters away to arrive at the moon's position. If, for example, you have the common 12-inch (one-foot) globe, you walk 30 feet; if you have an 8-inch globe, you walk 20 feet; and so on.

Start the moon journey by walking away from the globe in a straight line. Measure the distance with a ruler or tape measure. Glance back at the globe every few steps of your trip. How large does the Earth appear to you from the moon position? Whether it is smaller or larger than you expected, this is the size of the Earth as viewed by the lunar astronauts.

Astronaut William A. Anders, commenting on the appearance of the Earth from space, said: "The Earth looks so tiny in the heavens that there were times during the Apollo 8 mission when I had trouble finding it."

If you do not have enough space to reach the moon position, you may shorten the distance. Remember, however, that the globe will then appear larger than the Earth does from the moon. In other respects the viewing will be about the same.

A MONTH ON THE MOON

You leave your "space ship"; now the moon is "moving" you through space. Two rules govern your excursion:

1. Circle to the right (counterclockwise) around the Earth, keeping your present distance.
2. Always keep your face toward the Earth.

Since you are confined to ground level, you cannot travel in the exact orbit of the moon in relation to the Earth and the sun. Such an orbit would require you at times to fly above the surface and at other times to dig below it. What follows, therefore, is not to be regarded as a perfect model of a moon orbit. Moreover, since reflected light from the sky and ground will illuminate the globe, the "dark" area out of sunlight will not be black, although it will be in shadow.

THE NEW EARTH

Start your trip (Fig. 26). Observe the changing light and dark areas on the globe as you go. Make your first stop where the side of the Earth you see has the least amount of light. Here you will be looking in the direction of the sun, but do not stare directly at the sun. Bend your head low and you may improve the viewing. If the sun is very low in the sky (near sunrise or sunset as suggested earlier), the part of the globe you see should be completely in shadow.

You are seeing the "new Earth" from the moon. (It would be more realistic to call it "no Earth" because the side of the Earth you see has no sunlight at this time.)

Moon
30 29 28 27 26 25 24 23 22 21 20 19 18 17 16 15 14 13 12 11 10 9 8 7 6 5 4 3 2 1
Earth

Fig. 25. The moon is 30 Earths away from the Earth.

Fig. 26. A view of the Earth from the moon.

FIRST QUARTER EARTH

As you resume your face-the-Earth circuit, a crescent of sunlight advances over the globe. Note that the ends of the crescent—the **horns**—extend toward the left. The crescent grows into a half-circle. When the right half of the part of the globe you see is in sunlight and the other half in darkness, you have reached the "first quarter Earth" position.

You have traveled one-quarter of the way around the circle from "new Earth." This part of the moon's trip would take about a week.

FULL EARTH

Circling counterclockwise from the first quarter, you observe that the lighted area of the globe continues to grow. When you reach the place in the orbit opposite the new Earth position, the sun is directly behind you. You are halfway around the circle after two weeks of moon travel. Sunlight fills all of the side of the globe facing you. This is the full Earth.

It is now night on the side of the moon where you are; the sun is on its other side. (Feel its heat on the back of your head!)

THIRD QUARTER AND RETURN TO NEW EARTH

On the way to the third quarter the lighted part of the observed Earth shrinks while the shadowed part grows. At the third quarter, the *left* half of the Earth is lighted. The right half is dark. You have traveled for about three weeks.

Continue now toward the new Earth position. The sunlit area shrinks until a crescent Earth is seen again, this time with its horns pointing in the opposite direction from the previous crescent.

Arriving at the new Earth viewpoint, you have completed one orbit. Take the whole trip again, this time without pausing, to see the changes in the Earth's appearance during this period of a month.

It is evident that to a viewer on the moon, the Earth passes through the same phases or changes as the moon does for us. Do you think that the same phase, such as the "full phase," occurs on the Earth and the moon at the same time?

THE EARTH ROTATES

Phases of the Earth as viewed from the moon account for the changing size and shape of its light and dark areas. A lunar observer could also detect another change by fixing attention on the Earth's geography.

To demonstrate this during one of your circuits around the Earth, a friend would have to turn the globe on its axis twenty-nine-and-a-half-times. From any but the new Earth position you could observe the passage of the Earth's continents and oceans from the sunlight of day to the darkness of night, or the opposite, from night to day.

Viewed from the real moon, details of our geography are obscured by clouds that generally cover more than one-half of the total surface of the Earth (Fig. 27). The most striking feature of our planet, shown first in photographs taken on space missions, is that it appears enveloped in a deep aquamarine blue. Earth is the only blue planet in the solar system.

The Earth reveals all of its surface—clouded or clear—to a viewer on the moon facing our planet. On the other hand, an observer on Earth sees little more than one-half of the moon's surface even over a period of time. This should not surprise you if you reflect on the way that you conducted your face-the-Earth orbit.

When you, representing the moon, circled the Earth, you always kept your face toward our planet. Your friend on Earth could see only your face, never your back. It was only with the aid of lunar satellites equipped with TV cameras that we were first able to see the other side of the moon.

The moon rotates once on its axis each time it circles the Earth.

THE EARTH'S SIZE

Although to the casual observer on the moon the Earth would appear to have the same diameter throughout the month, an astronomer there would detect a difference. Why? The path you just followed was a circle with a radius of about thirty Earths. The moon's real path, however, is an **ellipse**.

At some places on the ellipse the moon is as close as twenty-eight Earths away; at other places it is almost thirty-two Earths away. Because of the varying distance, the Earth appears slightly smaller or larger at different times. You will find Earth-moon distances in Earth Facts on page 59.

Fig. 27. The Earth reveals all of its geography to a viewer on the moon if our planet is not obscured by clouds.

Chapter VII-One Square Inch of Sunlight

A beam of light originating in the excited atoms of the sun starts its space journey. Eight minutes and 93 million miles later it reaches the surface of the Earth. Here the solar radiation produces a wide range of effects. Some of the rays illuminate our planet and make it possible for us to see what is going on. Such visible light rays are also part of the source of energy for the food-making process in green plants. Ultraviolet rays cause sunburning and tanning, and produce vitamin D in your skin. Infrared rays, sometimes called heat rays, warm the Earth and make it habitable.

In the vastness of space, only a small part of the sun's radiant energy can fall on the Earth. The portion of it that does arrive here is distributed among the 197 million square miles of our planet's surface—but not in equal shares. Why this is so, and why different places are hot, cold, and everything in between, is revealed by the space-mounted globe.

NIGHT AND DAY

Allow the globe to warm up in sunlight for ten or fifteen minutes. Touch the night side of the globe with the palm of your hand, or your wrist, and then touch the sunlit side. It is not surprising that the sunlit side is warm and the dark side cooler. Nor is it difficult to understand why days are generally warmer than nights. Feel the part of the globe that has early morning. Then, glide your hand from west to east. Touch down first on the area directly under the sun—now having noon—and then on to the area having late afternoon. Make a brief detour to the place where you live. How do all compare?

TAKING THE GLOBE'S TEMPERATURE

Check your skin's sensitivity to temperature differences on the globe with a thermometer. Here is one way; perhaps you can invent others.

1. Determine the air temperature with the thermometer held in the shade in the area where your globe is set up. Make a record of it.
2. Place the thermometer face down on the spot of the globe to be measured, with its bulb held as closely as possible to the globe (Fig. 28). (The back of the thermometer frame should shield the bulb from the sun.)
3. Hold the thermometer in place for one minute. (You may vary this; but use the same time for each test.) Read the temperature at the end of the period of contact and make a record of it.
4. Allow the thermometer to come back to air temperature—in the shade—and go on to your next test.

This is admittedly a crude procedure, but it should indicate significant differences in the temperature in various parts of the globe.

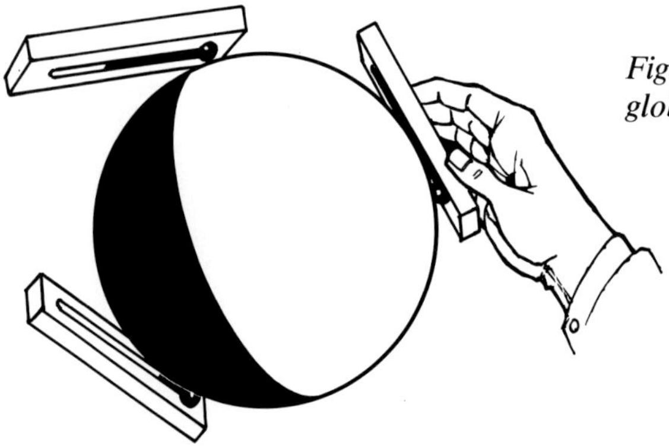

Fig. 28. Taking the globe's temperature.

FROM POLE TO POLE

Now move your hand from south to north on the sunlit part of your globe. Begin near the South Pole, touch down at the equator, move on to the place where you live, and then to the North Polar region. Can you feel the difference? Again, use the thermometer to check your sense impressions. Before reading on, try to guess why different parts of the globe have different temperatures—even when all are in the same sunlight.

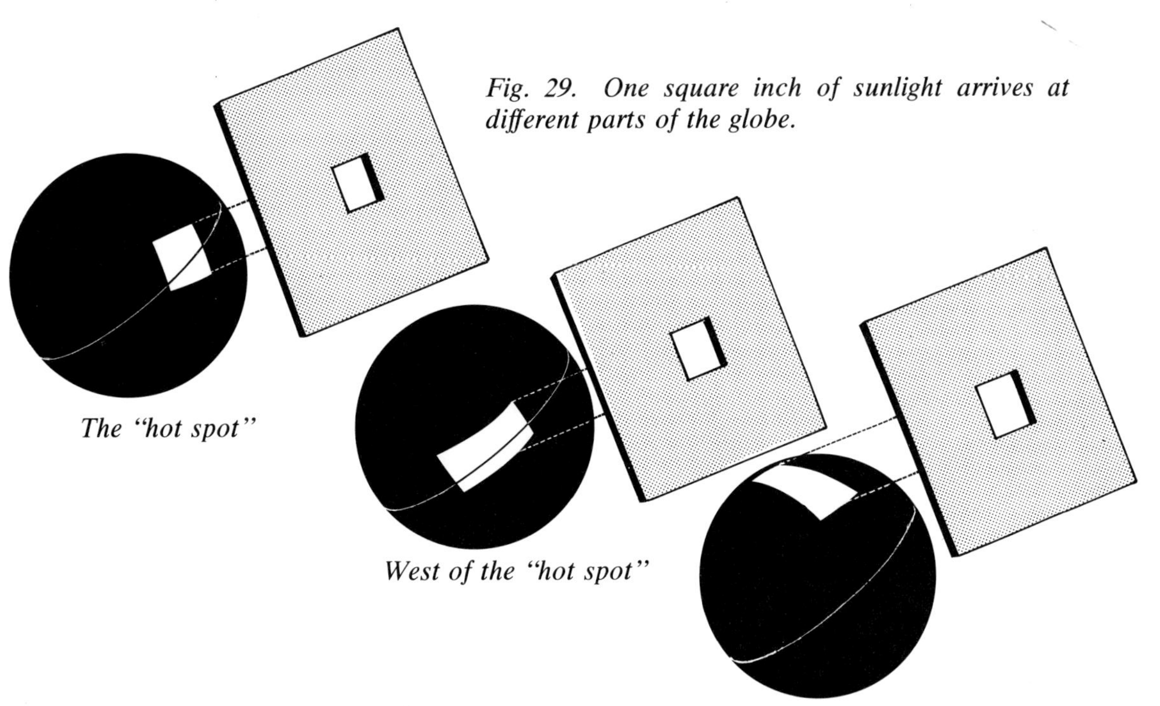

Fig. 29. One square inch of sunlight arrives at different parts of the globe.

FINDING THE HOT SPOT

Experiment with a tiny part of the sun's energy, just one square inch of it. To do this cut a one-inch square in an 8-×-11-inch cardboard as shown in the illustration.

Hold the cardboard facing the sun about six or seven inches from the center of the sunlit part of the globe (Fig. 29). To make sure that the cardboard is facing the sun directly, tilt it until the largest *square* possible is seen on that part of the globe.

Use a ruler to measure the square of sunlight on the globe. It should be one inch on each edge, or one square inch in area—the same size as the hole that you cut in the cardboard. If the patch of sunlight is rectangular, move the cardboard left or right, up or down, until it is as square as you can make it.

Exactly where on the globe is this square at this moment? Locate the city, country, or ocean in the center of the square. *It is interesting to note that in doing this you have found the part of the Earth that at this moment is facing the sun directly.* It is called the **subsolar point**; we might name it the "hot spot." People living at this location could say: The sun is exactly overhead now. And you, with your space view of the Earth, could add: The hot spot is the exact center of the daylight hemisphere.

Now move the cardboard toward the east, so that the sunlight falls near the sunset line. Keep it about the same distance from the globe as before but turn it so that it faces the sun directly. Again check to see that the spot of light is as large as possible. Is it still a square? Is it as bright as the hot spot? Measure it and enter the information about it on a chart similar to the partly completed one shown here.

In the same way, move the square inch of sunlight so that it falls as far west, then south, then north as possible, but stay within the sunlit area of the globe.

ONE SQUARE INCH OF SUNLIGHT

Place: Seaview, New York Date: Sept. 8, 1974				Time: 10:30 A.M. Air Temperature: 75° F	
Part of sunlit globe	Country or ocean	Latitude and longitude of center of spot	Shape of spot	Area of spot	Brightness
Center (Hot spot)	Colombia	5° north latitude 75° west longitude (approximate)	1" × 1"	1 square inch	very bright
Eastern edge					
Western edge					
Southern edge					
Northern edge	Greenland	70° north latitude 45° west longitude (approximate)	1" × 2"	2 square inches	dimmer

NORTH AND SOUTH OF THE HOT SPOT

From the evidence presented in the chart (which is a record of observations made on a 12-inch globe at the time, date, and place indicated), one square inch of sunlight illuminates one square inch of that part of the globe that faces the sun directly. In Greenland, however, where the sun's rays strike on a slant, one square inch of solar radiation is spread over two square inches of surface. As a consequence an area in Greenland (on the globe) at this time receives less heat than an equal area in Colombia. For the same reason other areas to the north or south of the "hot spot" receive less heat.

Because your globe is only a mini-earth, differences in the strength of the sun's radiation on its curved surface are not as great as they would be on the Earth itself.

EAST AND WEST OF THE HOT SPOT

In the early morning or late afternoon when the sun is near the horizon, a place on Earth receives less sunlight than at noon. Again slant makes the difference. You might expect, therefore, that noon would be the hottest time of day. This is not so, since the heat "piles up" hour after hour from sunrise on, so that the warmest time of day is usually in the midafternoon.

Evidently, the shape of the Earth and its spinning have much to do with how hot it gets on any day. In Chapter 10 you can investigate how the orbiting of our planet around the sun contributes to the changing temperatures that we experience during the year.

Eventually the heat that the Earth receives from the sun is radiated back to space. How can we trap and use more of it before it leaves our planet? Solar homes, solar water heaters, and solar batteries represent some of the efforts made to capture solar radiation. But this resource has scarcely been tapped: The energy that the United States receives from the sun in *two hours* is equal to that produced by our present fuel consumption in a whole year! If we can devise more efficient ways of using the vast supply of solar energy, we may be able to meet the world's ever increasing energy needs (Fig. 30).

Fig. 30. A proposed solar heat collector in a satellite orbiting the Earth at a height of 22,300 miles (35,680 kilometers). Solar cells on a 5 × 5 mile panel intercept solar energy twenty-four hours a day and convert it into electrical energy which is then beamed to a receiving station on Earth.

Chapter VIII-The Earth is a Clock

The Earth is a clock that never needs winding. So, too, is your globe, if you use it as a sundial. The first sundial was probably discovered by people who observed how the shadow of a tree moved during the day. In the early morning the shadow was long and pointed toward the west, in the direction opposite from the sun. As the morning wore on, the shadow shortened and turned toward the east. At noon, when the sun was most nearly overhead, the shadow was shortest. From noon to sunset the shadow lengthened as it continued its eastward swing around the tree.

The length of the shadow and its direction gave ancient people a notion of the time of day, long before the idea of hours came. Later a stick was deliberately thrust into the ground to serve as a shadow maker, and a circle of stones was placed around it to mark the shadow's position. Much later, when it was decided to divide the day into twenty-four hours, the stones may have been separated in such a way that the shadow moved from stone to stone in the course of an hour.

In this way, the two important parts of the sundial were invented: the shadow maker, called the **gnomon** (Greek for "one who knows"), and the surface where the shadow fell, with markings to show the passage of time, the **dial** (Latin for "day"). All future sundials had these two essential parts.

Much later it was found that if the gnomon was tilted at an angle that made it parallel to the axis of the Earth, the task of determining where the hour lines should be was simplified. You will recall that the axis of the space-mounted globe is also parallel to the Earth's axis.

A SUNDIAL IN YOUR HOME TOWN

How do you think you might use your globe as a sundial? One way is to set up a shadow stick on it at your hometown location, tilted so that it is parallel to the Earth's axis. The illustration shows how. For a gnomon, use a paper clip partly opened and straightened. Place it on the globe so that the straightened part (which is to serve as the gnomon) is over the place where you live. Hold it there with the thumb of one hand over the unopened part of the clip. To adjust the gnomon, look at the place where you live from the east or west side of the globe. Look at the North and South poles of the globe and estimate which way the axis runs through these poles. Bend the free end of the clip with your other hand so that it runs in the same direction as the globe's axis (Fig. 31). As a check, shift your view so that you look at the clip from above, directly over your home place. The gnomon should point northward—in the same direction as a nearby meridian. Fasten the base of the clip to your globe with "Magic" transparent tape. Your gnomon is now in place.

For a dial, cut out a circular piece of paper about two inches in diameter. Make a pinhole at its center and attach four strips of transparent tape to it. Run the dial down the gnomon and fasten it to the globe with the tape.

If the sun is shining, the shadow of the gnomon should be seen on the dial. Check with your watch, and when it reaches the hour, go over the shadow line with a pen from the center to about a quarter-inch from the edge of the dial. Mark the time there and indicate whether it is A.M. or P.M. If you start early in the morning and mark each hour, you will have a complete sundial by the end of the day. However, if you start in the afternoon, you can finish the job the next day.

From now on, you don't need a watch—at least not during the sunny hours. Moreover, the sun clock never needs winding—the Earth keeps it going.

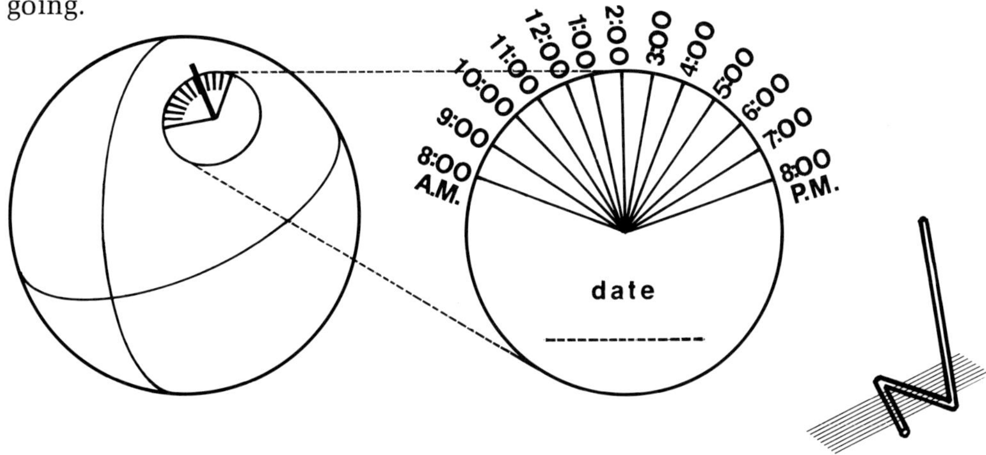

Fig. 31. A clip and a circle of paper convert your globe into a sundial.

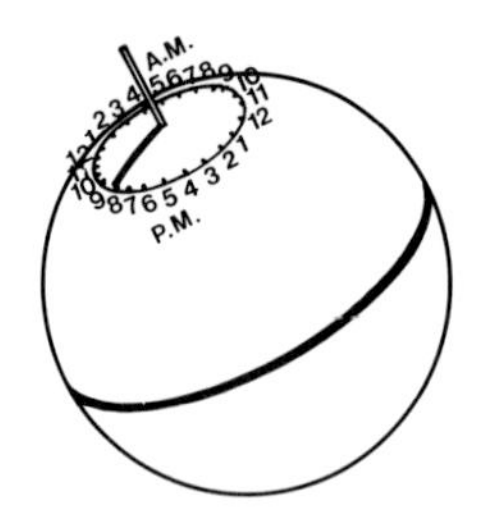

Fig. 32. A polar sundial is easily made by extending either pole.

A SUNDIAL AT THE POLE

Imagine that you want to make a shadow-stick sundial at the North Pole (Fig. 32). You fly there when this region has twenty-four hours of daylight—during the northern summer. What is the correct angle of the shadow stick? Directly *up*, since up is the direction of the Earth's axis there. You pound a pointed steel rod into the ice and watch its shadow. In what direction does it point? It must point south, no matter what time it is. Do you know why?

The polar ice around the steel rod will be your dial. You chop a groove in the ice under the shadow line. Every hour, using your watch, you chop a new groove. At the end of twenty-four hours, there will be twenty-four grooves, equally spaced, like the spokes of a wheel, and 15 degrees apart.

Now you have a sundial that tells time twenty-four hours a day but for only half of the year. (See drawing on page 32.) Of course, you could fly to the South Pole when darkness overtakes the North Pole.

As you watch the shadow at the North Pole, you note that it moves clockwise around the rod—in the opposite direction from that of the Earth itself. How do you think the shadow would move at the South Pole? (Fig. 33).

To set up a polar sundial without flying to a pole, erect a gnomon at either the North or South Pole of your globe. Which pole you use depends on the season of the year. In the summer months of the northern half of the Earth, the North Pole is the place; during the winter months, the South Pole.

Make your gnomon as you did before. Erecting it is easy, since it is just an extension of the Earth's axis, as shown in the illustration.

Prepare your dial in advance, drawing the hour marks for the full twenty-four-hour period. Indicate the A.M. and P.M. hours as shown. Note that if you are making a North Pole sundial the numbers go clockwise; for the South Pole, counterclockwise.

To set the time on the polar sundial, simply turn it until the gnomon's shadow falls on the correct hour (check with your watch). From now on it will serve as a sundial.

Alas, it cannot tell the time during *your* night because the Earth itself blocks the light of the sun from your Earth-based globe. In this respect, the globe is not perfect.

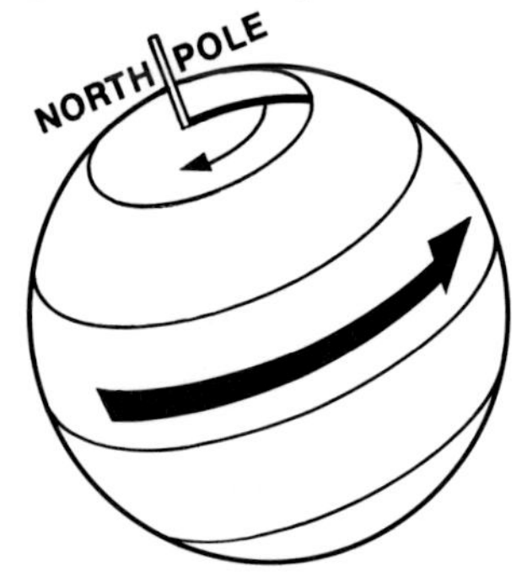

Fig. 33. Experiment with your globe to see how shadows move at the poles.

Chapter IX - What's Your Speed?

We are not inquiring about how fast you pedal a bike, drive a car, or fly a plane when we ask the question: What's your speed? Rather, we are thinking of your speed when you are sunbathing or star-gazing—when you appear to be motionless. What we are asking is—how fast is the Earth spinning you around?

The globe has the answer. With it you can determine your speed and also that of people anywhere on our planet. First, use the globe on its manufactured mounting and axis, if it has one—not in its free space position. To get the feel of the Earth's speed in different places, put a finger of one hand on the equator, and a finger of the other hand directly north of the first, near the North Pole. Now turn the globe on its axis from west to east (counterclockwise), halfway around. Which finger has traveled the greater distance? Do this again, with your arms extended and your head tilted back as far as possible, and watch your fingers and your hands.

Of course, the finger on the equator moves through a greater distance—but both fingers take the same time for the trip. It is easy to see that places on the equator move faster than those north or south of the equator. The farther from the equator, the slower the speed (Figs. 34 and 35).

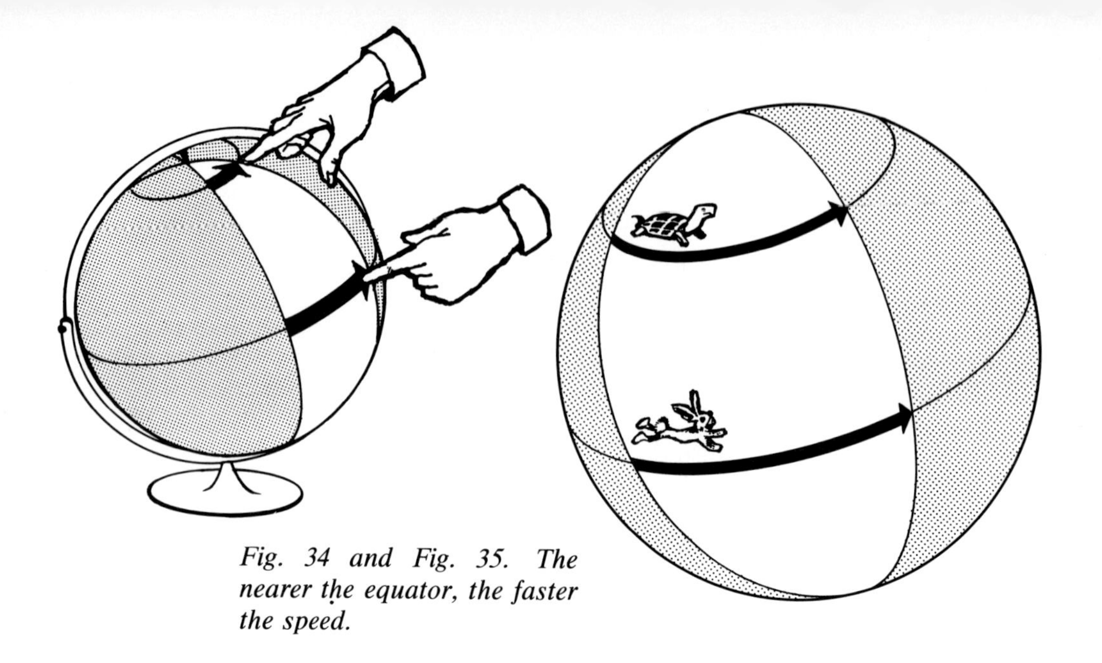

Fig. 34 and Fig. 35. The nearer the equator, the faster the speed.

FIFTEEN DEGREES AN HOUR

Just how fast are you traveling? Now place your globe in sunlight in the space position—your home town on top, your meridian in a north-south direction. Consider first the equator's speed. Mark an "X" on the equator where the sunset line crosses it, and note the time. One hour later, mark the new position of the sunset line. How many degrees difference is there between the two marks? (See the longitude figures printed along the equator.) Your answer should be close to 15°. This is the difference between any two meridians if your globe is marked with meridians 15° apart. See if the globe turns as much in a second and third hour.

The space-positioned globe, then, demonstrates how locations on the equator move at the rate of 15° in one hour (Fig. 36). In twenty-four hours, the Earth's equator will turn through twenty-four of these 15° arcs—a total of

Fig. 36. All places on Earth move the rate of 15° an hour.

360°, or one full turn. We know from the measurements and calculations of scientists that the circumference of the Earth at the equator is, in round numbers, 25,000 miles, or about 40,000 kilometers. This means that any point on the Earth's equator travels this distance in twenty-four hours. In one hour, the distance covered is about 1,040 miles, or about 1,670 kilometers. In short, the speed of a place on the equator is 1,040 miles, or 1,670 kilometers, an hour.

To find the speed of the Earth at the equator you had to know:

1. The distance traveled.
2. The time required to travel the distance.

WHERE YOU LIVE

How fast is the Earth spinning *you* around? You have to know only the distance it carries you for any period of time you choose.

As you found, places on the Earth's equator travel 15° an hour. So, too, do all places on Earth. This is apparent when you consider that the Earth is one sphere, and that all of its surface parts must spin together 360° a day or 15° an hour.

However, not all 15° arcs are equal! With a ruler or string measure the distance between two meridians 15° apart at the equator and then at the 60° north parallel. The first is twice as long as the second.

To find the distance *you* travel in an hour, it is necessary to find the length of the 15° arc on the parallel of your home town. Here is one way of finding this. (Perhaps you can invent others.)

Obtain a thin, transparent, flexible plastic ruler. Hold the ruler by its ends and lay it along the parallel that passes through your home town so that the edge with the numerals—inches or centimeters—follows it as closely as possible (Fig. 37). To do this it will be necessary to:

1. Hold the ruler upside down so that the numbers are on the bottom.
2. Bend the ruler to fit the curve of the parallel.
3. Tilt the ruler so that its edge lies as closely as possible to the surface of the globe.

If your place is not on a parallel that is printed on the globe (the chances are that it is not), use a nearby parallel to guide you. Measure the distance between two meridians 15° apart on your parallel. For Butte, Montana, for example, on a twelve-inch globe, the length is 1 3/32 inches or 2.8 centimeters.

To change globe inches into Earth miles, use the scale given on your globe. On one twelve-inch globe, the manufacturer states that one inch represents 660.5 miles. Similarly, on it one centimeter should represent 418.5 kilometers.

In one hour (according to the above measurements) Butte travels 660.5 × 1 3/32 miles, roughly 722 miles; or 418.5 × 2.8, roughly 1,172 kilometers. (Incidentally, this mathematical operation suggests one advantage of the metric system: It is easier to multiply by 2.8 than by 1 3/32.) Butte's speed is about 70 percent that of Singapore, which is located near the equator.

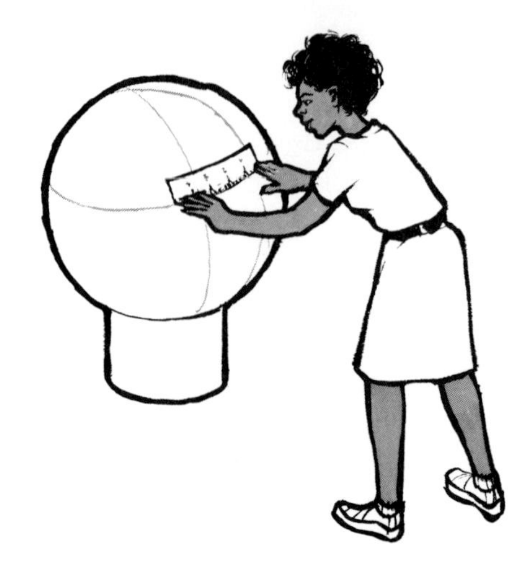

Fig. 37. To find the distance you travel in an hour, measure the length of a 15° arc on the parallel of your home town.

This "measure and multiply" method may be used to find the speed of any place on Earth.

At the Poles, the ends of the Earth's axis, the speed is zero, since there is no circling around the axis. However, an igloo on the Pole would make a complete turn once every twenty-four hours, counterclockwise at the North Pole, clockwise at the South Pole.

It seems strange that we do not feel the difference in speed when we change our latitude. At Anchorage, Alaska, for example, which is located on the 60° north parallel, the speed of rotation is only half that of the equator. The reason that we do not notice anything unusual is that everything around us in either place is moving at our speed. This is something like the experience of flying in a jet plane. Once the flying altitude is reached, one feels motionless, unless one sees a cloud nearby or a plane.

CURIOUS EFFECTS

The fact that the Earth's surface turns at different speeds at different latitudes makes for some interesting consequences. Imagine that an airplane starting out from Quito, Ecuador, near the equator, is "aimed" directly at Pittsburgh, Pennsylvania (Fig. 38). Assume that there are no winds that shift the plane from its course and that, once en route, the pilot does nothing to change the original heading or the altitude of the plane. (Of course, no plane is flown this way.)

When the plane is on the ground *before takeoff* it is moving, as is everything else at the equator, toward the east at the speed of 1,040 miles per hour. When it takes off and heads north, it is still sliding eastward with the Earth at that speed. If you were on the plane you would not be aware of the eastward drift because everything on the surface of the Earth and in the atmosphere above is turning eastward at that same speed.

As the plane flies north to higher latitudes, the Earth beneath, as we found before, progressively turns more slowly: at the 30° parallel (the latitude of Tallahassee, Florida), the Earth's eastward speed is only 900 miles an hour. What will happen to the plane, which still has its eastward equator speed of 1,040 miles an hour?

The plane drifts eastward with respect to the Earth below although it is still headed in the same direction, as shown in the illustration. At the 40° parallel that passes near Pittsburgh the Earth's speed is only about 800 miles per hour.

As a result of this drift from west to east, the plane crosses the 40th parallel east of Pittsburgh although the plane was "aimed" at Pittsburgh at the beginning of the flight.

This west to east motion is known as the *Coriolus effect,* after the 19th century French mathematician who first described it. As we have seen, it influences objects moving north in the Northern Hemisphere. What do you think will happen to objects flying south in the Northern Hemisphere or objects moving north or south in the Southern Hemisphere?

The Coriolus effect also influences the direction of the winds moving north or south in the Earth's atmosphere, deflecting them to the right of the direction they start from in the Northern Hemisphere, and to the left in the Southern Hemisphere.

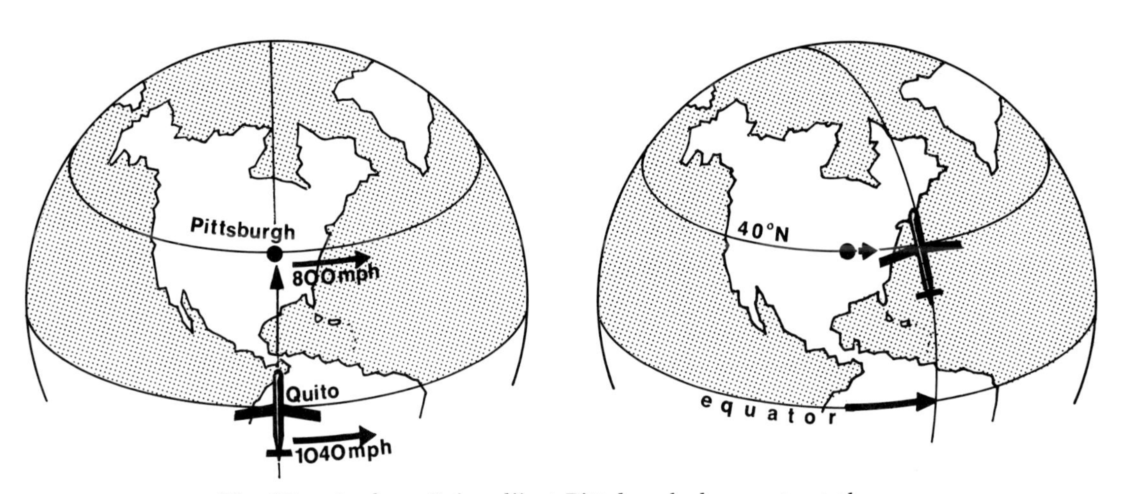

Fig. 38. A plane "aimed" at Pittsburgh does not get there.

MANY MOTIONS

The rotation of the Earth is only one of its many motions that carry you through space. While spinning, the Earth also circles around the sun in its yearly orbit at a speed of about 18.5 miles, or 29.8 kilometers a second. The sun, with its family of planets, including the Earth, is moving among the stars inside our Milky Way Galaxy. Our galaxy as a whole, is turning like a pinwheel and at the same time is moving among the other galaxies in the universe.

Chapter X-How Long the Day?

Sunrise and sunset border the sunlit half of the Earth. These boundary lines on your space-positioned globe can be used to measure the number of hours of daylight for any place on Earth, any day of the year. All that is necessary is to determine just where sunrise and sunset cross the parallel of the town or city in which you are interested.

COUNT THE HOURS

Look at the numbers printed on the equator for each of the meridians, the half-circles that run north-south from pole to pole. Begin with 0°, which marks the prime meridian that passes through Greenwich, England. Turn the globe slowly on its axis to the right from west to east. This is the way the Earth spins, taking twenty-four hours for a complete rotation.

On most globes, the marked meridians are spaced 15° apart. As you found in the previous chapter, 15° of turning means that one hour has elapsed. In one rotation of the globe, twenty-four meridian lines pass by. Twenty-four meridians, each 15° apart, make your globe into a twenty-four-hour clock. Continue turning the Earth on its axis and count the hours as the meridians pass by. Meridians, then, not only measure east-west distance; they also can be used to mark the hours (Fig. 39).

This is true no matter how many degrees apart the meridians on a globe are, since a degree is a measure of the fractional part of a circle. If your globe had only twelve meridians, each would be 30° apart—in order to make 360° in all. In one hour such a globe (turned by the Earth, of course) would spin only one-half of the distance between the two meridians, or 15°. Again, 15° of turning means that one hour has elapsed.

Some globes have each of the 15° spaces subdivided into single degrees on the equator. The passing of one degree means that four minutes (1/15th of 60 minutes) have gone by.

This is true both at the equator, where the meridians are widely spaced, and near the Poles, where they are crowded together.

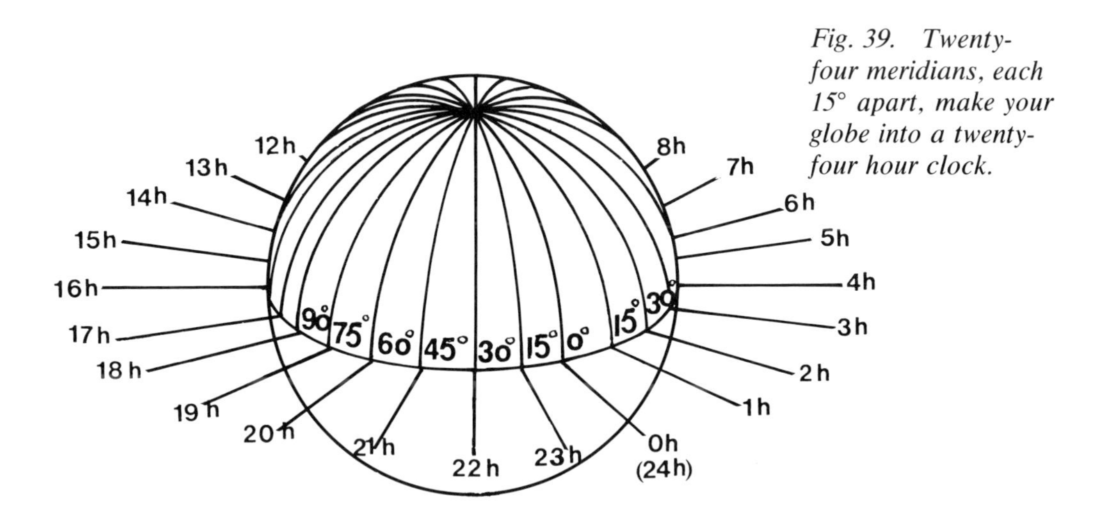

Fig. 39. Twenty-four meridians, each 15° apart, make your globe into a twenty-four hour clock.

SUNRISE TO SUNSET ON THE EQUATOR

To know how long a day is for a particular place we measure the distance in degrees along the parallel that runs through that place from the sunrise to the sunset line.

Tilt and turn your globe until it is in the correct space position in sunlight. How long is the day for a place near the equator, such as Singapore? An easy way to determine this is simply to count the number of the 15° meridian spaces along the equator from the sunset to the sunrise line. It may be necessary for you to include parts of spaces near the ends of this parallel to get the correct total. For example, it might be one-third of the space near the sunset line plus eleven full spaces plus two-thirds of a space near the sunrise line, a total of twelve full spaces. Twelve of these spaces also means twelve hours from sunrise to sunset—a twelve-hour day (Fig. 40).

It so happens that *all* places on the equator have a twelve-hour day every day of the year. A twelve-hour day also means a twelve-hour night. Does your space-positioned globe agree?

Fig. 40. Count the daylight hours along your parallel. The number of 15° meridian spaces between the sunset and sunrise lines equals the number of hours of daylight.

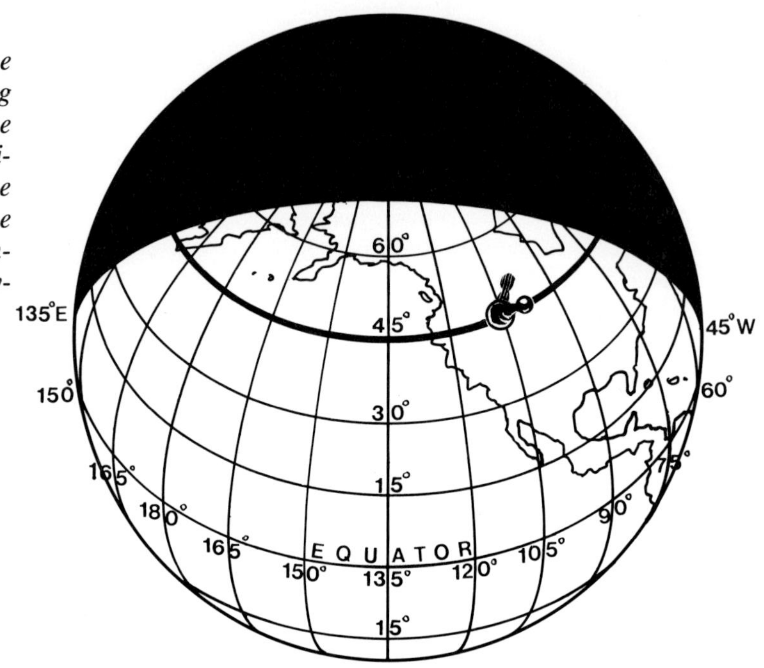

YOUR DAY

Now look at the place where you live. If it is on a parallel printed on the globe, your job is a bit easier than if you have to imagine this line. In either case, hold a pencil on your parallel so that it points toward the center of the globe. In this position slide it eastward to the point where it meets the sunset line. That is where its shadow should disappear. Make a mark there. Now run your pencil westward to the point where your parallel meets the sunrise line and make a mark there. Then count the number of meridian spaces on your parallel from sunrise to sunset (Fig. 40). Change these into hours, remembering that one space, or 15°, equals one hour and that 1/15 of a space, or one degree, equals four minutes. When you have found the answer, compare it with the one you compute from the sunrise and sunset hours given in the local newspaper for that day. Do they agree? Can you think of any reasons why there may be a difference? Whatever you find true for the length of day for your place is true for every place around the Earth on your parallel.

THE CIRCLE OF ILLUMINATION

There is a simple way to mark the entire sunrise and sunset lines on your globe at this time. In this way, you will be able to measure the hours of daylight *everywhere.* All you need for this is a long rubber band, one that can stretch easily around the globe (Fig. 41). It may help to prestretch the band between two "holders" such as two drawer pulls. Instead of a rubber band you may use a thin elastic string, sold in notion stores and hobby shops. The band or the string should fit loosely, so that it can be manipulated easily.

Look at your space-positioned globe in sunlight. Stretch the rubber band around the sunrise-sunset line. If you do this carefully, it will form a great circle around the globe, the **circle of illumination**. Step back, look, and adjust the rubber band until you are satisfied that it is in the right place. To shift it, hold one part of the rubber band while you roll another part of it into place—but do not move the globe out of its space position.

If you have done your work carefully, there should be twelve 15° spaces between the east and west part of the rubber bands on the equator; in other words, the circle of illumination should cut the equator into two equal halves.

To find the number of hours of daylight for any place, count the number of 15° meridian spaces for its parallel as you did before. This time, however, the rubber band will make it easier to see where any parallel is cut by the circle of illumination. Also, once the rubber band is snapped into place, you may pick up the globe and handle it at your leisure. As before, change the spaces and parts of spaces into hours and minutes. Select a number of cities around the globe and in the Northern and Southern Hemispheres.

If the city is on the nighttime side of the Earth at the time of your study, the same rule applies. Count the hours between the sunrise and sunset lines on its parallel on the sunrise side of the globe. Or, if you wish, count the nighttime hours on the night side and subtract from twenty-four.

It is interesting to find out how long the day is "down under," in Sydney, Australia, for example, when you are enjoying a long summer day, on July 4.

In Chapter 7 you experimented with "one inch of sunlight" and found that different parts of the Earth receive different amounts of sunlight. This happened because the angle at which the sun's rays struck the Earth's surface varied: the more direct strikes produce more heat than do the slanting ones. Now you have found another Earth-sun relationship that influences the amount of heat received: the length of day.

Long days and direct rays make summer.
Short days and slanting rays make winter.

There are four days each year when the length of day and night produce what we might call special Earth days, or Earth holidays. In the next chapter you will discover how this happens.

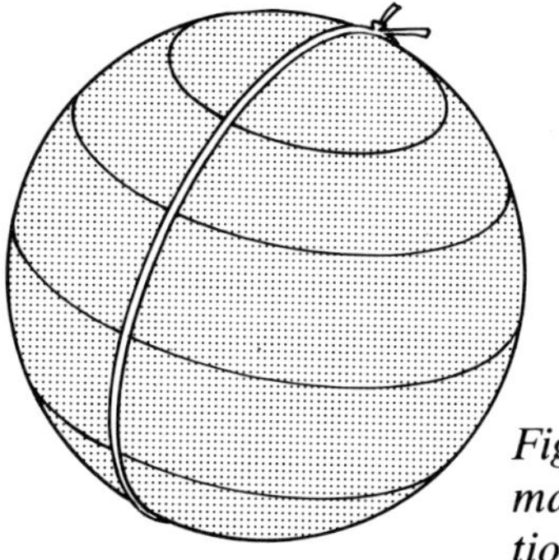

Fig. 41. An elastic band marks the circle of illumination and helps you find the number of hours of daylight anywhere.

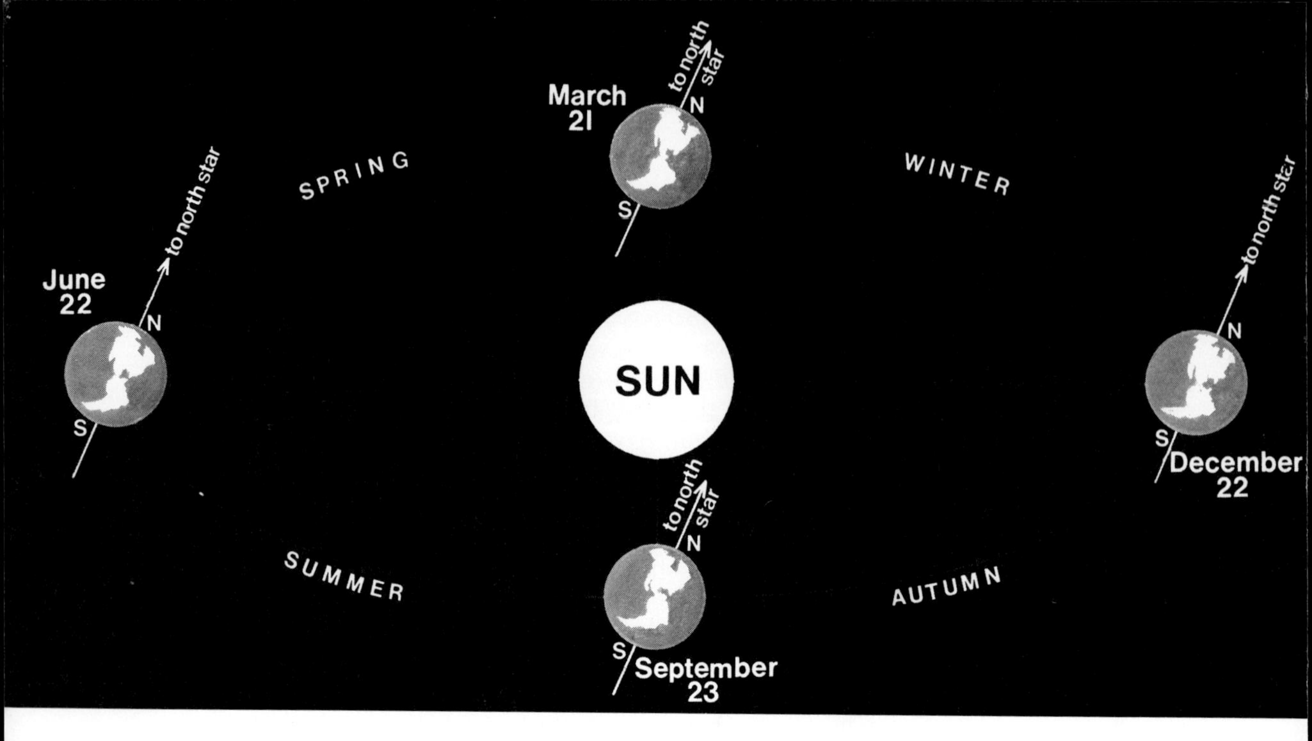

Chapter XI-Earth Takes a Holiday

Turned by the Earth, your space globe reveals what is happening on our planet on any day. At the same time that the Earth is spinning on its axis, it is also orbiting around the sun. Are signs of its annual circuit detected by the globe?

Look at your space-positioned globe on four days that we might call "Earth holidays." Customarily, they are known as the days that start the seasons. In the Northern Hemisphere we mark the beginning of spring on March 20th or 21st, summer on June 21st or 22nd, autumn on September 22nd or 23rd, and winter on December 21st or 22nd. Why are alternate calendar dates given for each of these?

The start of the seasons is not fixed by an Act of Congress or by any governmental body anywhere. Rather, it is determined by the actual instant when the Earth reaches each of four positions in its annual orbit of the sun. These would be the same each year—on our calendars and clocks—except for the fact that a complete circuit of the Earth around the sun requires about a quarter of a day more than the normal calendar year of 365 days. So, for example, the beginning of spring occurs about six hours later each year on our clocks for the three "normal" calendar years, after which the added day in leap year brings it back. The total result is to make spring's start range between March 20th and 21st. The other seasons start on alternate dates for the same reason.

It is difficult to see much of a difference in the appearance of the space-positioned globe from one day to the next. But if you keep a record—a photograph, sketch, or written account—of where the sunrise-sunset line circles the globe at intervals of a month or two, the change is easily detected. Another way is to mark the circle of illumination at noon on each of these days with a pencil—or to stretch an elastic around the globe, as suggested in the previous chapter.

Although ancient people thought that the sun—not the Earth—moved, the four special days were known to them because they were based on their accurate observations of the sun in the sky throughout the year. What is so "special" about these days?

DAY EQUALS NIGHT

We will start with the beginning of spring, March 20th or 21st. Now, as in any other time of the year, half the globe is in daylight, half in darkness. However, on this day the great circle of sunrise and sunset cuts **exactly through both poles**. The sunrise and sunset lines follow the meridians precisely from north to south. It looks as if the sun and your globe were made for each other: hour after hour the circle of light appears to pivot around the poles and swing around the meridians (Fig. 42).

How long is the day at this time? Find the length of day where you live by counting the number of spaces between the meridians from the sunrise to the sunset lines on the circle of latitude that passes through your town. Each space, on most globes, is 15° across.

No matter where you live, on this first day of spring there should be twelve of these 15° spaces between the sunrise and sunset lines—twelve hours of daylight; evidently all over the Earth except at the poles there are twelve hours of day and twelve hours of night.

The term **equinox** is given to this holiday. It comes from the Latin and means "equal night." Equal, that is, to day.

An equinox appears on one other day, September 22nd or 23rd, when the Earth in its orbit has moved 180° around the sun. Everything we have found out about the spring equinox applies to the autumn equinox. Check it on your globe and find out whether you agree.

Before leaving the equinox (it doesn't matter much if you observe it several days before or after the exact date), see how shadows fall on the Earth.

Place a shadow maker such as a pencil on the equator so that it points to the center of the globe (Fig. 43). It casts a shadow that falls exactly along the equator, no matter where you place it on this circle. The shadow extends directly east or west, except at the point directly under the sun, where there is no shadow at all. To a person at the equator on this day, this means that the sun rises exactly in the east, passes directly overhead at noon, and sets exactly in the west.

On the equinox days, then, the sun is seen directly overhead on the equator. Use your one-inch square (see Chapter 7) to see if the subsolar point—the "hot spot"—is indeed on the equator.

Try your pencil shadow maker elsewhere on the globe. There the shadow falls true east and west only near the sunset and sunrise lines. Glance at the position of the sun in the sky during the day—but do not stare at it. At the equinox the sun rises exactly in the east and sets exactly in the west, as seen from all places on the Earth.

We noted before that the only exception to the twelve-hour day, twelve-hour night cycle occurs at the Poles. What happens there? Use your shadow maker to find out. Does it help? It doesn't because the shadow does not fall on the globe at the time of an equinox—it extends out into space! To a person at either Pole the sun is exactly on the horizon all day, making a complete circle in twenty-four hours. You might expect to see half of the sun all day long!

Fig. 42. At the beginning of spring the circle of light cuts exactly through both poles.

SUN: STAND STILL!

The two other Earth holidays are the beginning of winter, December 21st or 22nd, and the beginning of summer, June 21st or 22nd. Consider first the onset of winter. Look at your space-positioned globe in sunlight at this time. The sunrise-sunset lines no longer run true north and south as they did during the spring and fall equinoxes. In the afternoon, note how the sunset line slants across North America from southeast to northwest.

The circle of light no longer cuts through the Poles. Rather, it just touches the Arctic Circle, which is 23½° south of the North Pole. On this first day of winter all places within the Arctic Circle have no sight of the sun for the entire twenty-four-hour period. At the same time the sun's rays shine down on the entire area included in the Antarctic Circle, which is 23½° north of the South Pole. For Antarctica this is the time of the midnight sun.

Fig. 43. Shadows at the time of the spring or fall equinox.

DAYLIGHT HOURS FOR DIFFERENT LATITUDES

DATE: (Dec. 22, 1975)		TIME: (4 P.M., E.S.T.)	
Latitude*	Degrees between sunrise and sunset	Hours of day	Hours of night
90° N (North Pole)			
66½° N (Arctic Circle)			
60° N			
(Your town's latitude**)			
30° N	(153° approximately)	(10 hrs., 4 min.***)	(13 hrs., 56 min.)
0°			
30° S			
60° S			
66½° S (Antarctic Circle)			
90° S (South Pole)			

*If the parallels of latitude on your globe are drawn at 10° intervals (90°, 80°, etc.) use those figures instead of those in the table.

**Insert in the proper place in this table.

***The times given in **The World Almanac** (Newspaper Enterprise Association, Inc., 1975.) are sunrise: 6:52, sunset: 17:05, which makes 10 hours and 13 minutes of daylight. It is very difficult to come close to this; expect an "error" of 5 or 10 minutes on a 12-inch globe.

Observe the areas within the Arctic and Antarctic Circles at different times during the day. Does the sun shine on the Artic? Does night come to the Antarctic? Watch these areas week after week until spring. How do they change?

Find out how long the day is for your place and other places on the globe. Do this as you did in the previous chapter by stretching an elastic band on the sunset-sunrise line so that it forms a great circle around the globe while it is in its space position. Then count the number of 15° spaces between the sunset-sunrise lines as you did during the equinox. Change this information into hours. (Remember that 15° is equivalent to one hour; one degree to four minutes). Determine the daylight hours for different latitudes, including your own. The sample table on this page may suggest a way of summarizing your observations and calculations.

When you look at your elastic-banded globe and when you examine your completed chart, it is evident that from north to south the number of hours of daylight increase while those of night decrease. Compare your findings with data given in a world almanac for the date.

It is also true that for the Northern Hemisphere as a whole this is the time of the shortest days and the longest nights of the year; the opposite is true for the southern half of the Earth.

Where is the sun directly overhead at noon at the beginning of winter? (Fig. 44). Use your "one square inch" or your pencil shadow maker to find the subsolar point on the globe. If you use the "one square inch," (see page 29) the subsolar point is in the center of the square inch of sunlight on the globe. If you use a pencil shadow maker (see page 20) it is the point of "no shadow." What is its latitude? It should be 23½° S, the Tropic of Capricorn.

June 21st or 22nd is the reverse of December 21st or 22nd. Both of these dates are known as **solstices** ("sol" means "sun"; "stice" means "stand"). To the ancients, the sun seemed to stand still before reversing its north-south motion in the sky. From June 21st to December 21st, the noonday sun appears to those in northern latitudes lower and lower in the sky with each succeeding day, appearing to "travel south" for the winter. Beginning with December 21st the sun starts to climb higher and higher until June 21st.

Conduct the same kind of investigation of the globe on the summer solstice as you did on the winter solstice.

WHY EQUINOXES AND SOLSTICES?

The equinoxes and the solstices occur because the Earth circles the sun once a year, with its axis always tilted (at an angle of 23½° from the perpendicular to the plane of the Earth's orbit around the sun) and "aimed" at a point in space near the North Star. These "holidays" mark four positions of the Earth in its orbit as shown in the drawing on page 44.

Locked in its proper space position, the globe reveals the passage of the seasons, but you don't have to wait for these special Earth holidays to make comparisons. Select your own days: the observations you make will fall in between those given here.

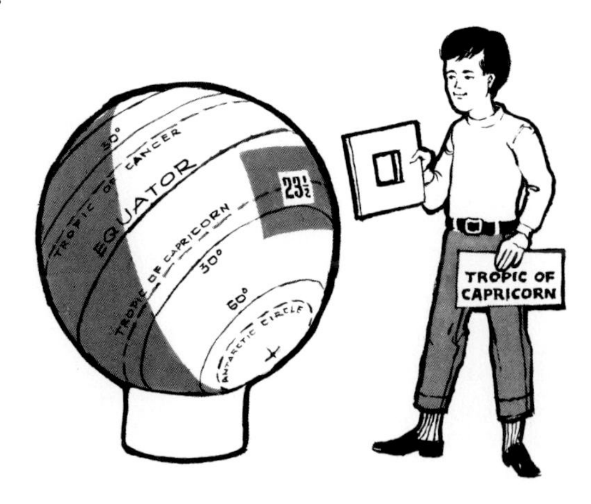

Fig. 44. During the winter solstice the "hot spot"—the subsolar point—is on the Tropic of Capricorn. The circle of light touches just below the Arctic Circle, leaving the area within it in darkness. At the same time, the area within the Antarctic Circle is in sunlight. It is now the Land of the Midnight Sun.

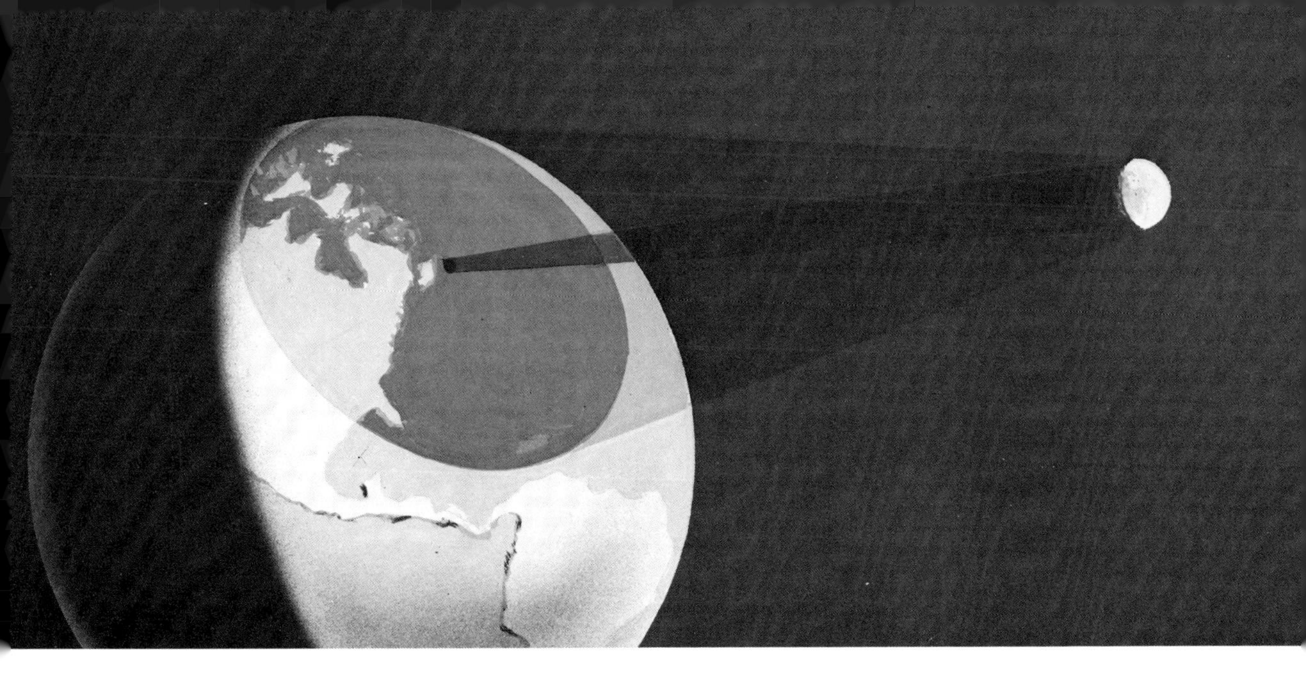

Chapter XII-Hide the Sun!

In its journey through space, Earth is accompanied by its natural satellite, the moon. Compared with the satellites of other planets, the moon is quite large, having a diameter one-quarter that of the Earth.

The moon circles the Earth about once every thirty days. You may recall from Chapter 6 that once during each circuit the moon is between the Earth and the sun, and then about two weeks later, it is opposite the sun. What happens when the sun, Earth, and moon are on a straight line in space?

For the purpose of this investigation a moon model out of proportion to the globe will be used (Fig. 45). A ball of modeling clay, a bead, a green pea, or any other sphere about one-fourth of an inch in diameter will do. Pierce the ball with a pin and mount it on a thin flexible wire about six inches long. A pipe cleaner is good, but first wet it and then twist it through your fingers to make it as thin as possible. Instead of a pipe cleaner you may use floral stem wire, used for wiring flower stems and sold in bundles in variety (five-and-ten-cent) stores. With the ball in place, make a small V-shaped loop at the other end of the wire. The loop will serve as a stand for the wire. The moon model is now ready for its space role.

PHASES OF THE MOON

First "try out" the moon model by moving it slowly around the space-mounted globe in a west-to-east, counterclockwise direction. Follow the line of the equator but keep the moon a foot or two away from the globe. (You may have difficulty in going all the way around, depending on how your globe is mounted.)

Look carefully at the moon model. How does the light on it change as it orbits the globe? Try to observe it from the viewpoint of an Earth person. To do this, hold the end of the wire at arm's length across the top of the globe, so that you are always on the opposite side of the globe from the model moon. Of course, this means that you have to circle around the globe yourself.

Keep the moon model in the sunlight at all times. If the globe blocks the light, hold the moon a bit higher or lower, or farther from the globe, if possible. If your body blocks the light, bend or twist to avoid this.

If you look carefully at the tiny moon model, you will see it go through its month: new moon (really "no moon"), first quarter, full moon, third quarter, new moon. These are the phases of the moon (Fig. 45). At all times, however, half of the moon is fully illuminated by the sun, but on Earth we do not always see this half.

HIDING THE SUN

Now bring the moon model closer to the Earth, about four-to-six inches away, and revolve it around the globe as before. When the moon is on the sunny side of the Earth its shadow races across it from west to east. When the moon is on the night side of the Earth, the shadow of the Earth darkens it (Fig.46).

On both of these occasions the sun is hidden:

It is hidden from the Earth by the moon, or
It is hidden from the moon by the Earth.

In the first instance, when the shadow of the moon falls on the Earth, we have an **eclipse of the sun**, but only for that part of the Earth in the moon's shadow.

In the second instance, when the shadow of the Earth falls on the moon, we have an **eclipse of the moon**. People all over the night half of the Earth see the moon slip into darkness.

Is there any relation between the phase that the moon is in and the occurrence of an eclipse? A study of your orbiting moon will suggest that an eclipse of the sun can take place only during the new-moon phase, and that an eclipse of the moon can occur only during the full-moon phase.

STAGING AN ECLIPSE

A most interesting "eclipse" of the sun may be staged by placing the moon model into orbit and watching the show. In this demonstration the sun is real and the space-mounted globe is a faithful reporter, but the moon model's orbit is only a fair imitation of a real one (Fig. 47).

Fig. 45. The moon in its first quarter phase.

Fig. 46. An eclipse of the sun is demonstrated.

Fasten the V-end of the wire with a strip of "Magic" transparent tape to the sunny side of the globe near the equator. Bend the wire toward the sun so that the moon model casts a shadow on the globe. To make an eclipse last as long as possible, your moon shadow should fall in the vicinity of the equator.

The moon model is now in orbit, but instead of revolving on its own, as the real moon does, it is carried by the globe. Its one-quarter-inch diameter produces a shadow whose size is in proportion to the one produced in a real eclipse. Its distance of five or six inches from the surface (of a twelve-inch globe) gives its moving shadow a speed that might occur in an eclipse.

What happens to the moon's shadow? Without disturbing the globe, outline the present position of the moon's shadow on it with a soft pencil. Make a record of the time in your notebook. At half-hour intervals repeat this until the shadow has disappeared from the surface of the globe.

The series of half-hour drawings on the globe is the track of the **eclipse path**, the places where the eclipse can be observed by Earth dwellers. Apparently, the sun is hidden by the moon only over a narrow strip of the Earth.

The eclipse ends as the moon's shadow slips off the Earth. Study your

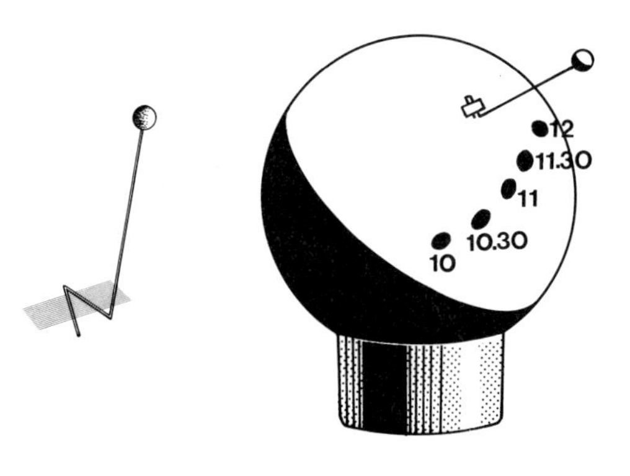

Fig. 47. An eclipse of the sun may be staged by placing a moon model in orbit. Watch the eclipse path until the shadow slips off the globe.

STAGED ECLIPSE OF THE SUN

DATE:	PLACE: (your home town)		
Time (in half hour intervals)	Location of Shadow		Shape of Shadow
	Latitude	Longitude	
9 A.M.			
9:30 A.M.			
10 A.M.			
10:30 A.M.			
11 A.M.			
11:30 A.M.			
12 noon			
12:30 P.M.			
1 P.M.			
1:30 P.M.			
etc.			

globe, copy the table shown here and fill it in with your observations. Then try to answer the following:

1. How many hours elapsed between the first and the last observations of the moon's shadow?
2. How long was the eclipse path? (Measure it with a string. Then lay the string along a ruler and count the inches or centimeters. Multiply by the scale factor to get the distance covered. On a twelve-inch globe it is about 660.5 miles to the inch, or 418.5 kilometers to the centimeter.)
3. What was the average speed of the eclipse across the Earth? (Miles per hour $= \frac{\text{miles}}{\text{hours}}$; kilometers per hour $= \frac{\text{kilometers}}{\text{hours}}$)
4. What was the shape of each of the observed eclipse shadows that you outlined? Did they change?
5. In what phase was the moon during the eclipse as viewed from the Earth?

On a second day, try to stage several eclipses in different parts of the Earth at the same time. Place a number of moons into orbit at different latitudes above and below the equator. Collect data, record it in a table, and answer the same five questions as above.

PUZZLING THOUGHTS

Puzzle 1. If the moon circles the Earth once a month, why don't we have an eclipse of the sun and of the moon every month?

To understand why we don't, hold a "moon" by the wire and circle it around the globe in sunlight as best you can at a distance of a few feet. Sometimes the shadow falls on the Earth; sometimes it misses. This also happens to the real moon because it is not always lined up exactly or almost exactly with the sun and the Earth.

Another way of looking at it is this: Picture the sun set in a depression in the center of a round table. The Earth moves around the edge of the table. The surface of the table represents the plane of the Earth's orbit. At the same time, the moon tags along with the Earth, circling it once a month. However, the moon's orbit around the Earth is not exactly on the level of the table: It is tilted so that at times the moon rises above it and at other times below it. At these times no eclipse of the sun can occur because the moon's shadow passes above or below the Earth. Similarly, no eclipse of the moon can occur because the Earth's shadow will not fall on the moon. To have an eclipse it is necessary:

1. For the moon to be in the right position in relation to the Earth and sun (new-moon position for eclipse of the sun; full-moon position for an eclipse of the moon), and
2. For the moon to be "on the table"—on the same plane as the Earth and the sun.

Puzzle 2. The moon's diameter is only about 1/400 that of the sun. How can it eclipse or hide the sun?

The sun is about 400 times as far away from the Earth as the moon. Both appear to an Earth viewer to be the same size—which means that if the moon is in line with the sun and the Earth it can obscure the sun completely (Fig. 48).

Our Earth and moon models do not lend themselves to a demonstration of partial eclipses. For this and other information about eclipses consult the books listed on page 60, or visit a local planetarium or astronomy club. Best of all, try to observe a real eclipse, total or partial. Consult a current almanac to find the dates and places of solar and lunar eclipses for the year.

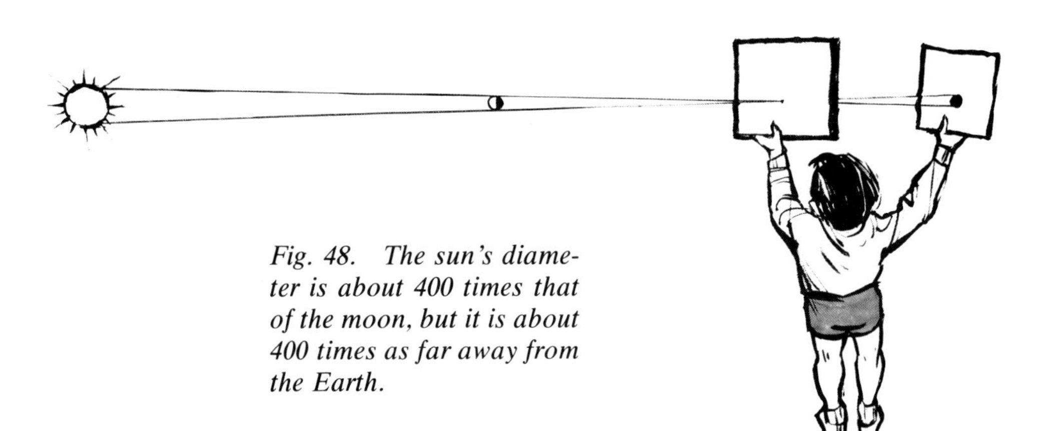

Fig. 48. The sun's diameter is about 400 times that of the moon, but it is about 400 times as far away from the Earth.

Chapter XIII - By the Light of the Moon

The moon, like the Earth, is a sunlit sphere in space. At the time of the full moon it turns all of its sunny side toward the Earth. In this phase the moon may be regarded as our planet's "nighttime sun." It rises in the east just about the time the sun disappears in the west. After spanning the sky the moon retires in the west as the sun comes up in the east.

What part of the Earth is illuminated by the moon? The globe, positioned properly on its pad, "works" in moonlight just as it does in sunlight to bring you news of what the nighttime side of the Earth is doing in space.

MOONLIGHT EXPEDITION

This investigation is limited to the full moon only. A clear night is essential, of course. Equip yourself with a flashlight, watch, compass, ruler, notebook, and soft pencil (or any other globe-marking device that you have pretested). A golf tee or chess piece to stand over your home town will help you position it "on top."

If you have a choice, a light-colored globe is preferred because it will reflect light better than a darker one. Set the globe up in its space position in a place as free as possible from house or street lights. The shadow box described on page 19 or a piece of cardboard will help shield the globe from interfering light.

The nighttime sky provides a guide for checking your "north." Find the Big Dipper and the two stars at the front of the bowl, as shown on page 8.

These are known as the "pointers." A line drawn through the pointers and extending about five times the distance between them will lead you to the Pole Star, known also as the North Star. Now sight along the axis of your globe. Does it point to the North Star?

FIRST IMPRESSIONS

Allow five or ten minutes for your eyes to accommodate to the darkness.

What part of the globe is in moonlight? How about the area around the North Pole? The South Pole?

Take a walk around the globe at a distance of ten or fifteen feet to see its light and dark sides. The light side of the globe is illuminated by moonlight. The opposite side is dark because it is in the shadow of the Earth. If your globe were hoisted into space, its "dark side" would be in sunlight—just as the Earth is.

Come close to the globe. Can you make out the continents, countries, and oceans now dimly illuminated by moonlight?

MOONRISE AND MOONSET

Where on Earth is the moon rising at this moment of viewing? Find the moonrise line in the west. You may have to turn your flashlight on to identify the places on the moonrise line.

Look to the east to locate the moonset line and to find the places near it. Which is closer to your home town—moonrise or moonset?

THE CIRCLE OF MOONLIGHT

The moonrise and moonset lines join to make a great circle around the Earth. Locate the circle on your globe and run your finger around it. Mark out the circle by making about a dozen points on its circumference. To do this, try the "disappearing-shadow" method that you used in sunlight. Slide a pencil close to where you think the circle is. Where the shadow disappears mark an "X." When you have completed the staking out of the circle make a note of the time (Fig. 49).

Look at the moon at this time and determine its general direction: north, northeast, east, southeast, south, southwest, west, or northwest.

Repeat the procedure a half- or a full hour later to locate the new position of the circle of moonlight on the globe and of the moon itself. This time mark an "O" at each point to distinguish it from the first marking. Later, when you take the globe indoors, you can study the circles at your leisure. But for the present, continue your moonlight adventure.

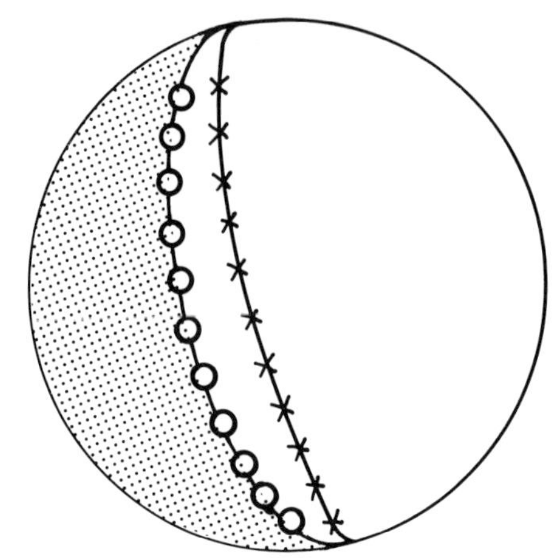

Fig. 49 The progress of the circle of moonlight.

HOW HIGH THE MOON?

Despite its apparent brilliance, the moon shines with less than 1/400,000 the brightness of the sun. Even if the entire night sky were packed with full moons, the total illumination would be less than one-fifth that of bright sunlight. Moonlight, which is reflected sunlight, differs from it in other ways: The *percentage* of the potentially harmful ultraviolet light (the kind that produces sunburn) and infrared light (heat rays) is much, much less. It is for this reason that the moon, unlike the sun, may be stared at without danger.

Watch the progress of the moon during the night. When the moon reaches its highest position in the sky—some time around midnight—point to it with an outstretched arm (Fig. 50). Thrust your other arm directly forward. Estimate the angle formed by your arms (or make a sketch of it). This angle, expressed in degrees, is called the altitude.

How does the moon's altitude at midnight compare with that of the noonday sun on the same or the next day?

Generally, the full moon is high in the sky at the time of the year when the sun is low—during the winter. It is low when the sun is high—during the summer. See if you can explain why this is so with the help of the illustration on page 44 which shows the Earth at different "stations" in its yearly orbit around the sun.

Fig. 50. The moon's altitude is measured in degrees.

HOW LARGE IS THE MOON?

Observe the moon in the early evening, when it is close to the horizon, and then later at night. Is there a real difference in the observed size—or is this some kind of an illusion? One way to find out is to measure the diameter of the moon in its low and high positions in the sky.

Hold a ruler in your hand at arm's length, with four fingers wrapped around it from behind and the thumb in front (Fig. 51). Move the ruler until its edge appears to cut the moon vertically in two halves. Adjust it so that its end lines up flush with the top of the moon. Then slide your thumb up the ruler until your nail is at the same level as the lower rim of the moon. Use your flashlight to see what the measurement is. How do the measurements of the high and the low moon compare? Does the moon change size during the night?

CIRCLE X AND CIRCLE O

When you take the globe indoors complete the circles that you marked out in moonlight. First stretch a long rubber band or an elastic string around the globe over the "X's." It should form a great circle, dividing the Earth into two equal halves. Does the half of the globe in moonlight include the Arctic area? The Antarctic area? Both? Does the moonrise line run north and south along a meridian or does it slice across the meridians?

Now stretch a second band around the "O's." How does the "O" circle compare with the "X" circle? Where do the two circles cross each other? Your answers to these questions will depend on the time of the year that you are observing your space-oriented globe.

UNEXPECTED DISCOVERIES

You have found how the globe, in moonlight or sunlight, can monitor planet Earth's journey through time and space. Many unexpected discoveries await you as you continue to use the globe as an instrument for conducting your own Earthwatch.

Fig. 51. Does the moon change in size during the night?

Appendix
Metric and Customary Units

Approximate Conversions from Metric to Customary Units

When you know	Multiply by	To find
Length		
millimeters	0.04	inches
centimeters	0.4	inches
meters	3.3	feet
kilometers	0.6	miles
Area		
square centimeters	0.16	square inches
square meters	10.8	square feet
square kilometers	0.4	square miles
Temperature (exact)		
Celsius	$9/5$ (then add 32)	Fahrenheit

Approximate Conversions from Customary to Metric Units

When you know	Multiply by	To find
Length		
inches	2.54 (exact)	centimeters
feet	30	centimeters
miles	1.6	kilometers
Area		
square inches	6.5	square centimeters
square feet	0.09	square meters
square miles	2.6	square kilometers
Temperature (exact)		
Fahrenheit	$5/9$ (after subtracting 32)	Celsius

Earth Facts

(Symbols: km=kilometers; mi=miles; sq km=square kilometers; sq mi=square miles; sec=seconds)

Distances	Metric Units	Customary Units
Diameter at the equator	12,756.3 km	7,926.4 mi
Diameter at the poles	12,713.5 km	7,899.8 mi
Circumference at the equator	40,075 km	24,901.5 mi
Distance from the sun		
greatest	151,800,000 km	94,500,000 mi
least	147,100,000 km	91,400,000 mi
average	149,500,000 km	92,900,000 mi
Distance from the moon		
greatest	406,686 km	252,710 mi
least	356,400 km	221,463 mi
average	384,403 km	238,857 mi
Length of orbit around the sun	1,100,699,276 km	683,942,855 mi
Areas		
Land area (29.2%)	148,940,000 sq km	57,506,000 sq mi
Water area (70.8%)	361,130,000 sq km	139,433,000 sq mi
Total area	510,070,000 sq km	196,939,000 sq mi
Velocities		
Of rotation at latitude 0° (equator)	1,675.99 km per hr	1,041.41 mi per hr
30°	1,451.29 km per hr	901.79 mi per hr
60°	839.32 km per hr	521.53 mi per hr
90° (poles)	0.00 km per hr	0.00 mi per hr
Of revolution in orbit around sun (average)	29.8 km per sec	18.5 mi per sec

Books about Space, Time and the Globe

Abell, George. **Exploration of the Universe.** New York: Holt, Rinehart and Winston, 1969. A college text on astronomy.

Asimov, Isaac. **The Clock We Live On.** New York: Abelard-Schuman, 1965. How astronomy, geography, and history have influenced our clocks and calendars.

Branley, Franklyn M. **The Earth: Planet Number Three.** New York: Thomas Y. Crowell, 1966.

——— **Experiments in Sky Watching.** New York: Thomas Y. Crowell, 1967.

——— **The Moon: Earth's Natural Satellite.** New York: Thomas Y. Crowell, 1972.
Observations, experiments, and information leading to an understanding of our planet.

Greenhood, David. **Mapping.** Chicago: University of Chicago Press, 1964. The meaning and making of maps.

Hirsch, S. Carl. **The Globe for the Space Age.** New York: Viking, 1963. How the globe helps us understand the Earth.

Joseph, Maron J. and Lippincott, Sara L. **Point to the Stars,** Revised Second Edition. New York: McGraw-Hill, 1977. How to locate the constellations and the stars.

Marshall, Roy K. **Sundials.** New York: Macmillan, 1963. History, design, and use of sundials.

Strahler, Arthur N. **The Earth Sciences.** New York: Harper and Row, 1971. Useful sections on the globe; changes resulting from the rotation and revolution of the Earth.

Sutton, Richard M. **The Physics of Space.** New York: Holt, Rinehart and Winston, 1965. Observations and experiments leading to an understanding of the application of the laws of physics to space.

Also consult almanacs for current Earth events and atlases for detailed maps.

Index

Page numbers of illustrations are in *italic*.